Introduction to Biophysical Chemistry

INTRODUCTION TO BIOPHYSICAL CHEMISTRY

Edited by:

Dr. Pratibha Singh and Dr. Varsha Srivastava

ARCLER PRESS

www.arclerpress.com

Introduction to Biophysical Chemistry

Dr. Pratibha Singh and Dr. Varsha Srivastava

Arcler Press
224 Shoreacres Road
Burlington, ON L7L 2H2
Canada
www.arclerpress.com
Email: orders@arclereducation.com

ISBN: 978-1-77469-866-2 (Hardcover)

ABOUT THE EDITORS

Dr. Pratibha Singh received Ph.D degree in Chemistry, and has research experience in Natural Dyes Extraction. She was Project associate in the Department of chemical engineering IT BHU, India, working on Water Sample Testing and Effluent Treatment. She also has work experience as a Scientific Officer in the field of Surface Chemistry. She has authored a number of national and international publications.

Dr. Varsha Srivastava received Ph.D. degree in Chemistry from the Department of Applied Chemistry IIT BHU. She has a research experience in the field of Polymer based Corrosion Inhibitor for Mild Steel in acidic media. She was also awarded with Fellowship UGC JRF under RFSMS scheme (Research Fellowship in Science for Meritorious students). She served as a Senior Lecturer (Associate Professor) in Babu Banarasi das University (BBD, NIIT) Lucknow. Some of her research work were published in the Journal of Applied Polymer Science, Journal of Applied Electrochemistry, and Corrosion Science.

TABLE OF CONTENTS

LIST OF FIGURES

SUMMARY

Biophysical chemistry is a multidisciplinary field that uses models of chemistry and physics to comprehend organic systems by interpreting the quantitative, energetic, structural, functional, and interactional context of the physical nature. In this introductory volume, the authors provide insight into the significance of biophysical chemistry as a developing area of science and the wide variety of strategies that have been applied to explain the related phenomena. All of these techniques cannot be covered in such a brief introduction because they have spectroscopic, electrochemical, thermal, and physiological elements. Rather, the volume discusses a few specific methods for reviewing biological molecules and how they function. A few of these procedures use environmental and semi-environmental factors as that for organic particles in the natural environment. The volume focuses on different methods that up to date are utilized in biophysical chemistry. Every method comes with a brief definition and is cited with proper reference for future research.

PREFACE

Due to the complex nature of biological systems, different classical and sophisticated methods have been used in the discipline of biophysical chemistry. Thus, in order to describe biological systems, mathematical, physical, and chemical methods all combine into one stream. As a developing field of study, biophysical chemistry is currently growing up fast with significant breakthroughs recently. In fact, three biophysical chemists were given the 2009 Nobel Prize in Chemistry for their research on X-ray ribosome diffraction. The key areas that attract biophysical chemists include; sub-atomic structure, function, dynamics and kinetics, relations and macromolecule thermodynamics which are situated in the cytoplasmic matter or cell membrane.

Overall, biophysical chemistry discusses the following topics: functions of biological molecules, cell effects if some biological molecules are dysfunctional, biological processes, molecules or particles involved in the process and their structure, duration of biological process and what energetics accompany the change.

As a discipline, biophysical chemistry concentrates on the quantitative and substantial aspects of matter. Through the intellectual perspective of biophysical chemistry, the authors examine biological content by keenly defining the basics of its physical context. The book provides readers with the benefit of using insights from the researchers to explore the complexities of biological systems. The basic philosophical context of biophysical chemistry is that the application of these disciplines to biological systems can prompt significant knowledge acquisition.

It can be challenging to explore biological factors and biophysical perspective without being proficient about the basic physical and chemical elements. While teaching and learning these essentials, both the educator and the student can become puzzled in the physical concepts and fail to understand the model of biophysical chemistry. In this book, the authors endeavor to provide the correct balance and simplify the concepts for all readers to grasp.

Biophysical Chemistry discusses the physical nature of macromolecules and the experimental strategies used to research them. Some of the topics covered include: basics of biological molecules; mass spectrometry, spectroscopy, and macromolecule hydrodynamics; biomolecular kinetics; and electrophoresis; and single-molecule methods. The digestible, pragmatic style captures the reader's attention with strong theoretical and experimental concepts. This book covers practical aspects for students studying chemical and physical sciences and will serve as a context for further research. The less complex approach to physical chemistry that has been presented here will be appreciated by students in different fields of biological sciences. With verified examples and a focus on independent learning, each chapter presents the fundamental ideas behind each topic in a clear and concise manner.

The book presents original works in the fields of physics and chemistry directly affecting organic phenomena. Quantitative analysis of the elements of biological macromolecules, biologically active particles, macromolecular groups, and cell elements on thermodynamics, kinetics, spatio-temporal structures, X-ray and NMR structural biology, including single-atom detection cover a significant part of the publication. Hypothetical and computational treatments of bio-macromolecular structures, macromolecular infrastructures, regulatory standards, and biological systems are additionally mentioned.

CHAPTER 1

Principles of Physical Chemistry (Basic Rules and Constants)

Contents

1.1. INTRODUCTION

Physical Chemistry is a branch of science that deals with the study of the properties of matter and the changes it undergoes when subjected to various forces. Physical Chemistry is also known as Physical Science. It deals with chemical reactions, thermodynamics, electrochemistry, mechanics, and kinetics. Physical Chemistry incorporates concepts from physics, chemistry, and mathematics in order to understand chemical reactions and their mechanisms. Physical chemistry is a very important branch of science because it has a direct impact on many other branches of science like biochemistry, biology, medicine, etc. The principles of physical chemistry are used by scientists to predict or explain certain phenomena such as chemical reaction or even nuclear fusion or fission (De Paula, 2011).

The study of physical chemistry begins with an exploration of atomic structure, where scientists delve into the intricate world of subatomic particles and their interactions. Atoms, the building blocks of matter, consist of a nucleus composed of protons and neutrons, surrounded by electrons orbiting in specific energy levels. By understanding the arrangement of these particles and their properties, physical chemists gain insights into the behavior and properties of different elements (Adamson, 2012).

One of the key concepts in physical chemistry is the periodic table, a powerful tool that organizes the elements based on their atomic number, electron configuration, and chemical properties. This arrangement reveals patterns and trends that allow us to predict and understand the behavior of elements and their compounds. From the periodic table, we can discern the electronegativity, ionization energy, and electron affinity of elements, which play a crucial role in chemical reactions and bonding.

Quantum mechanics, a foundational theory in physics, forms the basis for understanding the behavior of atoms and molecules. In the microscopic world, particles exhibit wave-particle duality and can only exist in discrete energy states. Quantum mechanics provides a mathematical framework to describe the behavior of these particles, including their wave functions, energy levels, and probabilities of occupying different states. This understanding is vital for comprehending the principles behind chemical bonding, molecular structure, and spectroscopy.

Intermolecular forces, the attractive and repulsive interactions between molecules, significantly influence the physical and chemical properties of substances. These forces dictate phenomena such as boiling and melting points, solubility, and viscosity. Understanding intermolecular forces allows us to explain why some substances are gases at room temperature while others are solids and how different substances interact to form solutions or precipitates (Adamson, 2012).

Chemical bonding lies at the heart of physical chemistry. It involves the sharing, transfer, or redistribution of electrons between atoms to form stable compounds. Covalent bonds result from the sharing of electron pairs, while ionic bonds arise from the transfer of electrons between atoms. Through these bonding mechanisms, molecules acquire unique shapes and properties. Molecular geometry, determined by the arrangement of atoms and lone pairs of electrons, influences the reactivity and physical behavior of compounds (De Paula, 2011).

The study of physical chemistry also encompasses the behavior of gases, where the ideal gas law plays a central role. This law, derived from the principles of thermodynamics, describes the relationship between the pressure, volume, temperature, and the number of moles of a gas. It provides a framework to understand gas behavior under different conditions, enabling scientists to predict and manipulate gases in various applications.

Stoichiometry, the quantitative study of chemical reactions, allows us to determine the amounts of reactants and products involved. By applying the principles of stoichiometry, chemists can calculate reaction yields, identify limiting reagents, and determine the theoretical and actual quantities of products obtained. This information is crucial for designing and optimizing chemical processes.

Acid-base concepts form another fundamental aspect of physical chemistry. The behavior of acids and bases can be explained by theories such as the Arrhenius, Bronsted-Lowry, and Lewis definitions. These theories describe the transfer of protons (H+) or electron pairs in acid-base reactions and help explain pH, acid dissociation constants, and buffer solutions. Acid-base reactions play a crucial role in biological systems, industrial processes, and environmental chemistry.

Thermochemistry, the study of heat energy changes in chemical reactions, is central to physical chemistry. The first law of thermodynamics, often called the law of energy conservation, states that energy can neither be created nor destroyed but can only be converted from one form to another. Thermochemistry allows us to quantify energy changes associated with chemical reactions, including enthalpy, heat capacity, and calorimetry. These concepts are vital for understanding the energy flow in chemical systems and designing efficient processes (Adamson, 2012). The concept of entropy, intimately tied to the second law of thermodynamics, characterizes the degree of disorder or randomness in a system. It provides insights into the direction and spontaneity of chemical reactions. The second law states that the total entropy of an isolated system always increases in a spontaneous process, highlighting the importance of energy dispersal and the tendency towards higher disorder.

Entropy plays a fundamental role in chemical equilibria, phase transitions, and the prediction of reaction feasibility (De Paula, 2011). Through the study of physical chemistry, scientists gain a deeper understanding of the fundamental rules and constants that govern the behavior of matter.

These principles, ranging from atomic structure to quantum mechanics, intermolecular forces, chemical bonding, and thermodynamics, provide a framework to interpret and explain chemical phenomena. With this knowledge, researchers can explore and design new materials, understand complex biological processes, develop sustainable energy solutions, and unravel the mysteries of the universe.

1.2. ATOMIC STRUCTURE AND THE PERIODIC TABLE

The atomic structure of an element is the way that its atoms are arranged in a certain pattern or combination of patterns. The periodic table is a chart that describes which elements are arranged in which patterns. Atoms are the smallest unit of an element.

Each atom has a nucleus and electrons that orbit it, making up the structure of an atom. Atoms can also combine to form molecules, which are made up of multiple atoms bonded together. The size of an atom is given by its atomic mass, which is equal to the atomic weight of its constituent elements. This reference table organizes all known chemical elements according to their atomic number, their electronic configurations and other physical properties (Figure 1.1).

Atoms, the building blocks of matter, are incredibly small and cannot be observed directly with the naked eye.

They are composed of three main subatomic particles: protons, neutrons and electrons. Protons carry a positive charge, neutrons are electrically neutral and electrons possess a negative charge. These particles are arranged in a unique manner within the atom, contributing to its overall structure and properties (Adamson, 2012).

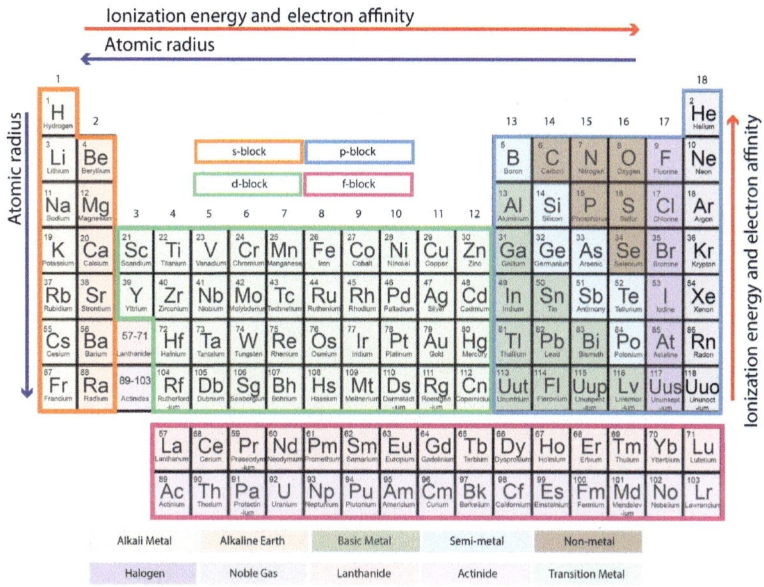

Figure 1.1. The periodic table.

Source: Harpp, K. (2018). Periodic Table. In: White, W.M. (eds) Encyclopedia of Geochemistry. Encyclopedia of Earth Sciences Series. Springer, Cham. https://doi.org/10.1007/978-3-319-39312-4_245

The central core of an atom is called the nucleus. It contains protons and neutrons tightly bound together. Protons, with their positive charge, determine the identity of an element. Each element is characterized by a specific number of protons in its nucleus, known as the atomic number. For example, hydrogen has one proton, helium has two protons and so on.

The total number of protons in an atom defines the element's place in the periodic table (De Paula, 2011).Neutrons, on the other hand, have no electrical charge. They add mass to the nucleus without significantly affecting the chemical properties of the atom.

The sum of protons and neutrons in the nucleus gives the atomic mass of an atom. Since the mass of an electron is negligible compared to that of a proton or neutron, it is usually not considered when calculating the atomic mass. Electrons, the negatively charged particles, orbit around the nucleus in specific energy levels or shells. These energy levels are commonly referred to as K, L, M and so on, with the K shell being the closest to the nucleus. Each energy level can hold a certain number of electrons. The K shell can hold a maximum of 2 electrons, the L shell can hold a maximum of 8 electrons and so on. The distribution of electrons in these energy levels determines the chemical behavior and reactivity of an atom.

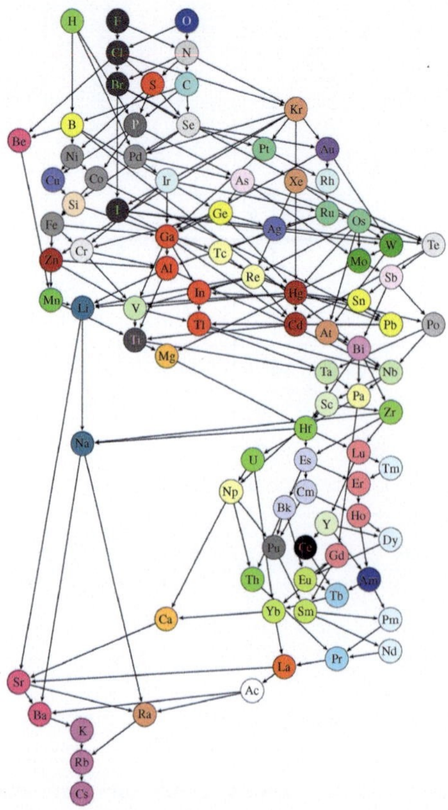

Figure 1.2. Formal structure of periodic system of elements.

Source: Leal Wilmer and Restrepo Guillermo 2019Formal structure of periodic system of elements. Proc. R. Soc. A. http://doi.org/10.1098/rspa.2018.0581

The arrangement of electrons in an atom follows a set of principles known as the Aufbau principle, the Pauli Exclusion Principle and Hund's rule (Figure 1.2). The Aufbau principle states that electrons fill the lowest energy levels first before moving to higher energy levels. The Pauli Exclusion Principle states that no two electrons in an atom can have the same set of quantum numbers, which includes their spin. Hund's rule states that electrons fill orbitals of the same energy level one at a time before pairing up (Adamson, 2012).

The behavior of electrons within an atom can be visualized using atomic orbitals. These orbitals are regions of space where there is a high probability of finding an electron. The most commonly known atomic orbitals are the s, p, d and f orbitals. The s orbitals are spherical in shape and can hold up to 2 electrons. The p orbitals are dumbbell-shaped and can hold up to 6 electrons. The d and f orbitals have more complex shapes and can hold up to 10 and 14 electrons, respectively.

The arrangement of elements based on their atomic number, electron configuration and chemical properties is facilitated by the periodic table. The periodic table is a tabular arrangement of elements organized in rows and columns. Elements with similar properties are grouped together in the same column, known as a group or a family. Elements in the same group exhibit similar chemical behaviors, valence electron configurations and often have analogous properties (De Paula, 2011).

The periodic table is divided into several blocks: s-block, p-block, d-block and f-block. The s-block consists of the first two groups (Group 1 and 2), which include elements such as hydrogen, lithium, sodium and potassium. These elements have their valence electrons in the s orbital. The p-block comprises groups 13 to 18 and includes elements like nitrogen, oxygen, carbon and chlorine. These elements have their valence electrons in the p orbital (Adamson, 2012).

The d-block consists of transition metals, which are found in the middle of the periodic table (groups 3 to 12). These elements have their valence electrons in both the s and d orbitals. Transition metals exhibit unique properties such as variable oxidation states and the ability to form complex ions. The f-block contains the inner transition metals, which are located at the bottom of the periodic table. These elements have their valence electrons in the f orbitals.

The periodic table also provides information about atomic mass, atomic symbol and other properties of elements. Elements are arranged in increasing atomic number from left to right and top to bottom. Periods, the rows of the periodic table, represent the number of energy levels or shells an atom possesses. The number of valence electrons, the outermost electrons involved in chemical bonding, increases from left to right across a period.

The periodic table has undergone various modifications and improvements since its inception. The modern periodic table is based on the work of Dmitri Mendeleev and Julius Lothar Meyer, who independently proposed organizing elements according to their properties. Mendeleev's periodic table was arranged in order of increasing atomic mass and predicted the existence and properties of yet-to-be-discovered elements. The periodic table we use today has evolved to incorporate additional information, such as the atomic number and electron configuration.

1.3. QUANTUM MECHANICS AND ITS RELEVANCE TO PHYSICAL CHEMISTRY

Quantum mechanics, a fundamental theory in physics, provides a powerful framework for understanding the behavior of matter at the atomic and subatomic levels. Its principles and mathematical formalism are essential for unraveling the mysteries of physical chemistry, where atoms, molecules and their interactions are explored. In this section, we will delve into the key concepts of quantum mechanics and examine their relevance to the field of physical chemistry. Quantum mechanics revolutionized our understanding of the microscopic world by introducing the concept of wave-particle duality (Figure 1.3). According to this principle, particles such as electrons and photons can exhibit both wave-like and particle-like properties. This duality implies that the behavior of particles is described not by classical mechanics but by mathematical wave functions (De Paula, 2011).

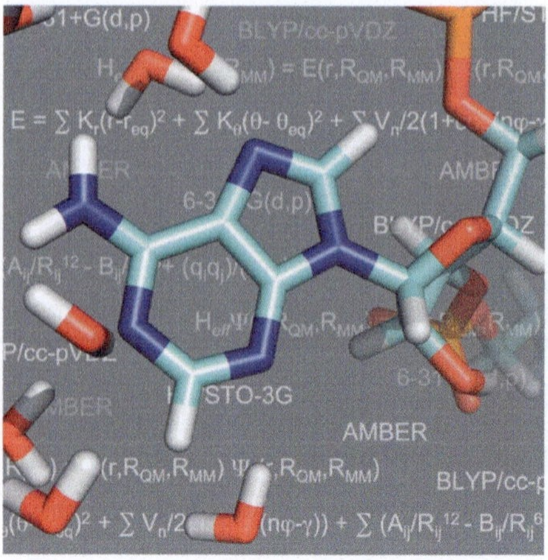

Figure 1.3. Molecular dynamics and quantum mechanics of RNA.

Source: Ditzler, Mark A. and Otyepka, Michal and Šponer, Jiřì and Walter, Nils G, (2010), Molecular Dynamics and Quantum Mechanics of RNA: Conformational and Chemical Change We Can Believe In, Accounts of Chemical Research, volume 43, pages 40–47, https://doi.org/10.1021/ar900093g

The wave function, denoted by the Greek letter Ψ (psi), provides a mathematical description of the quantum state of a particle. It contains all the information about the particle's position, momentum and energy. The absolute square of the wave function, |Ψ|^2, gives the probability density of finding the particle at a particular position. In other words, the wave function determines the likelihood of observing a particle in a specific state (Adamson, 2012).

One of the central equations in quantum mechanics is the Schrödinger equation. This partial differential equation describes how the wave function of a particle evolves over time. It incorporates the particle's kinetic energy, potential energy and the effects of external forces. Solving the Schrödinger equation provides the wave function and, consequently, the complete description of the particle's behavior.

The quantization of energy is a key aspect of quantum mechanics. According to the theory, energy levels in quantum systems are discrete rather than continuous. This means that certain allowed energy states are separated by specific energy intervals. The quantization of energy is particularly evident in atomic systems, where electrons occupy distinct energy levels or orbitals. Electrons in atoms and molecules are governed by a set of rules known as quantum numbers. These numbers describe various characteristics of electrons, such as their energy, orbital shape and spin. The principal quantum number (n) denotes the energy level or shell in which an electron resides. The azimuthal quantum number (l) specifies the shape of the electron's orbital, such as s, p, d, or f orbitals. The magnetic quantum number (m) defines the orientation of the orbital in space. Finally, the spin quantum number (s) describes the intrinsic angular momentum of the electron (De Paula, 2011).

The concept of electron orbitals is crucial in understanding chemical bonding and molecular structure. Orbitals are regions of space where there is a high probability of finding an electron. Each orbital can hold a specific number of electrons. The s orbital, for instance, is spherical and can hold a maximum of 2 electrons. The p orbitals are dumbbell-shaped and can hold up to 6 electrons (2 electrons in each of the three p orbitals).

Chemical bonding, a fundamental concept in physical chemistry, is deeply rooted in quantum mechanics. The formation of chemical bonds involves the sharing, transfer, or redistribution of electrons between atoms. The behavior of electrons in bonding is described by molecular orbitals, which are constructed by combining atomic orbitals.

In molecular orbitals, electrons are not confined to specific atoms but are delocalized over the entire molecule. The combination of atomic orbitals leads to the formation of bonding and antibonding molecular orbitals. Bonding orbitals have lower energy and contribute to the stability of the molecule, while antibonding orbitals have higher energy and weaken the bond. The number of bonding and antibonding molecular orbitals depends on the number of atomic orbitals involved in the combination.

The quantum mechanical model of chemical bonding, known as valence bond theory, explains the formation of covalent bonds. It describes bonding as the overlap of atomic orbitals, where electrons are shared between atoms. The strength and stability of covalent bonds depend on the degree of orbital overlap and the resulting electron density between the atoms (De Paula, 2011).

Another influential theory in chemical bonding is molecular orbital theory. This theory considers the entire molecule as a single entity and describes bonding in terms of molecular orbitals. By solving the Schrödinger equation for the molecule, molecular orbitals are obtained, which can be occupied by electrons. The molecular orbital theory allows for a more comprehensive understanding of bonding in molecules and predicts properties such as bond order, bond length and magnetic properties.

Spectroscopy, the study of the interaction between matter and electromagnetic radiation, is another area where quantum mechanics plays a crucial role. Spectroscopic techniques, such as infrared spectroscopy, UV-Vis spectroscopy and nuclear magnetic resonance (NMR) spectroscopy, provide valuable insights into the structure, composition and dynamics of molecules. Quantum mechanics explains how molecules interact with light and how energy is absorbed, emitted, or scattered during these interactions. Spectroscopy relies on the quantization of energy levels in atoms and molecules, which leads to the absorption or emission of discrete amounts of energy. By analyzing the absorption or emission spectra, scientists can identify functional groups, determine molecular structures and investigate the electronic and vibrational properties of molecules (Adamson, 2012).

1.4. INTERMOLECULAR FORCES AND THEIR EFFECTS ON CHEMICAL PROPERTIES

Intermolecular forces play a crucial role in determining the physical and chemical properties of substances. These forces are the attractive and repulsive interactions that exist between molecules and are responsible for phenomena such as boiling and melting points, solubility and viscosity. Understanding the different types of intermolecular forces and their effects on chemical properties is vital for comprehending the behavior of substances and designing new materials. In this section, we will explore the various intermolecular forces and their implications in physical chemistry. The three main types of intermolecular forces are dispersion forces (also known as London dispersion forces), dipole-dipole forces and hydrogen bonding. These forces arise due to the presence of charged particles (either permanent or temporary) in molecules, leading to attractions or repulsions between neighboring molecules.

- Dispersion forces are the weakest intermolecular forces and exist in all molecules, regardless of their polarity. These forces result from temporary fluctuations in electron distribution, creating instantaneous dipoles in the molecules. The temporary dipole induces a corresponding dipole in neighboring molecules, leading to attractive forces. Dispersion forces increase with increasing molecular size and surface area. Substances with stronger dispersion forces tend to have higher boiling and melting points and greater cohesion between molecules (De Paula, 2011).
- Dipole-dipole forces arise in polar molecules where there is a permanent separation of charge. In these molecules, the electronegativity difference between atoms creates a permanent dipole, with one end being more electronegative (partially negative) and the other end being less electronegative (partially positive). The positive end of one molecule attracts the negative end of a neighboring molecule, resulting in dipole-dipole interactions. Dipole-dipole forces are stronger than dispersion forces but weaker than hydrogen bonding (Adamson, 2012).
- Hydrogen bonding is a specific type of dipole-dipole interaction that occurs when a hydrogen atom is bonded to a highly electronegative atom (such as oxygen, nitrogen, or fluorine) and is also attracted to another electronegative atom in a different molecule. The electronegative atom (typically oxygen or nitrogen) acts as the hydrogen bond acceptor, while the hydrogen atom acts as the hydrogen bond donor. Hydrogen bonding is responsible for many unique properties and phenomena in substances, such as the high boiling points of water and the DNA double helix structure.

The strength of intermolecular forces influences several important chemical properties. One of these properties is boiling point. Substances with stronger intermolecular forces require more energy to overcome these forces and transition from a liquid to a gas state. Consequently, substances with stronger dispersion forces, dipole-dipole forces, or hydrogen bonding tend to have higher boiling points. For example, water, with its extensive hydrogen bonding network, has a higher boiling point compared to similar-sized molecules with only dispersion or dipole-dipole forces.

Melting point, similarly to boiling point, is affected by intermolecular forces. Substances with stronger intermolecular forces require more energy to break these forces and transition from a solid to a liquid state. Consequently, substances with stronger intermolecular forces have higher melting points. For example, ionic compounds, with their strong electrostatic attractions between ions, have high melting points compared to covalent compounds with weaker intermolecular forces. Solubility, the ability of a substance to dissolve in another, is influenced by intermolecular forces. Like dissolves like, meaning that substances with similar intermolecular forces tend to be soluble in each other. Polar substances tend to dissolve in polar solvents through dipole-dipole interactions or hydrogen bonding. Nonpolar substances, on the other hand, tend to dissolve in nonpolar solvents through dispersion forces. This principle guides the solubility of substances in various solvents and is essential in areas such as organic chemistry and drug formulation (De Paula, 2011).

Viscosity, the resistance of a substance to flow, is also affected by intermolecular forces. Substances with stronger intermolecular forces have higher viscosities because the molecules are held together more tightly, impeding their ability to move past one another. For example, substances such as honey or molasses, with strong intermolecular forces, exhibit high viscosity, whereas substances such as water, with weaker intermolecular forces, have lower viscosity.

Surface tension, the force that holds the surface of a liquid together, is another property influenced by intermolecular forces. Molecules at the surface of a liquid experience stronger attractive forces with neighboring molecules than with the surrounding air or another medium. This cohesive force creates a surface tension that allows liquids to form droplets or exhibit capillary action. Substances with stronger intermolecular forces have higher surface tension. Understanding intermolecular forces is also vital for comprehending phase changes, such as evaporation, condensation and sublimation. In these processes, intermolecular forces are either overcome or formed, leading to the transition between different states of matter (solid, liquid, or gas) (Adamson, 2012).

1.5. CHEMICAL BONDING AND MOLECULAR GEOMETRY

Chemical bonding is the process by which atoms are held together to form molecules and compounds. It is a fundamental concept in chemistry that governs the behavior and properties of substances. The type of chemical bond formed between atoms determines the structure, stability and reactivity of molecules. Additionally, molecular geometry, which describes the arrangement of atoms in a molecule, plays a crucial role in determining its physical and chemical properties. In this section, we will explore the different types of chemical bonding and the significance of molecular geometry in understanding and predicting molecular behavior.

There are three main types of chemical bonds: covalent bonds, ionic bonds and metallic bonds. Each type of bond arises from the interaction of valence electrons between atoms, leading to the formation of a stable molecular structure.

Covalent bonds occur when atoms share one or more pairs of electrons. This sharing of electrons allows each atom to achieve a more stable electron configuration, resembling the noble gas configuration. Covalent bonds are typically formed between nonmetal atoms and can occur between identical atoms (as in diatomic molecules) or different atoms (as in polyatomic molecules). The strength of a covalent bond depends on the number of shared electrons and the distance between the nuclei of the bonded atoms (De Paula, 2011).

Ionic bonds, on the other hand, are formed between atoms with a significant difference in electronegativity. In an ionic bond, one atom donates electrons to another atom, resulting in the formation of positively charged cations and negatively charged anions. These oppositely charged ions are held together by electrostatic attractions. Ionic bonds are typically observed in compounds composed of metals and nonmetals. Ionic compounds often form crystal lattice structures and tend to have high melting and boiling points due to the strong electrostatic attractions between ions.

Metallic bonds are specific to metals and are responsible for their unique properties. In metallic bonding, the valence electrons of metal atoms are delocalized, meaning they are free to move throughout the metal lattice. This delocalization creates a "sea" of electrons that hold the metal atoms together. Metallic bonds contribute to properties such as high electrical and thermal conductivity, malleability and ductility in metals (Adamson, 2012).

The concept of molecular geometry is crucial in understanding the physical and chemical properties of molecules. Molecular geometry refers to the three-dimensional arrangement of atoms in a molecule and is determined by the arrangement of electron pairs around the central atom. The arrangement of atoms affects the distribution of charge, the polarity of the molecule and the spatial orientation of chemical bonds.

One of the fundamental theories used to predict molecular geometry is the valence shell electron pair repulsion (VSEPR) theory. According to this theory, electron pairs (both bonding and nonbonding) around a central atom repel each other and tend to be as far apart as possible. This repulsion determines the overall shape of the molecule. The VSEPR theory allows us to predict the molecular geometry based on the number of bonding and nonbonding electron pairs around the central atom. The VSEPR theory identifies several basic molecular geometries. These include linear, trigonal planar, tetrahedral, trigonal bipyramidal and octahedral geometries, among others. For example, a molecule with two electron pairs around the central atom adopts a linear geometry, while a molecule with three electron pairs adopts a trigonal planar geometry. A molecule with four electron pairs adopts a tetrahedral geometry and so on. The VSEPR theory provides a simplified but effective model for predicting molecular shapes and understanding the arrangement of atoms in a molecule (De Paula, 2011).

The polarity of a molecule, which is determined by the presence of polar bonds and the molecular geometry, is also crucial in understanding its behavior. Polar bonds arise when there is an unequal sharing of electrons between atoms due to differences in electronegativity. The electronegativity difference creates a separation of charge within the bond, resulting in partial positive and partial negative charges. The polarity of a molecule depends on both the polarity of its individual bonds and the overall molecular geometry.

In a molecule with polar bonds, the molecular geometry may result in a net dipole moment. For example, a molecule with a linear geometry and two polar bonds can be nonpolar if the polar bonds are arranged symmetrically, canceling out each other's effects. On the other hand, a molecule with a bent geometry and two polar bonds can be polar if the polar bonds are arranged asymmetrically, leading to a net dipole moment. The polarity of a molecule affects its solubility, intermolecular interactions and chemical reactivity (Adamson, 2012).

In addition to molecular geometry, hybridization is another important concept in chemical bonding. Hybridization involves the mixing of atomic orbitals to form new hybrid orbitals that are more suitable for bonding. The concept of hybridization allows us to explain the observed shapes of molecules and the types of bonds they form. For example, sp^3 hybridization occurs when carbon atoms form four sigma bonds in a tetrahedral geometry, as seen in methane (CH_4).

The strength and stability of chemical bonds significantly impact the physical and chemical properties of substances. Covalent bonds, with their sharing of electrons, tend to be relatively strong and require substantial energy to break. Consequently, substances with covalent bonds often have high melting and boiling points.

Ionic bonds, on the other hand, are generally stronger than covalent bonds and involve the complete transfer of electrons. Substances with ionic bonds tend to have higher melting points, as the electrostatic attractions between ions are strong. Metallic bonds, characterized by the delocalization of electrons, contribute to the unique properties of metals, such as electrical conductivity and malleability.

Understanding chemical bonding and molecular geometry allows scientists to predict and interpret the behavior and properties of molecules and compounds. It provides insights into the strength of chemical bonds, the shape of molecules, the polarity of compounds and their reactivity. This knowledge is essential for designing new materials, understanding biological processes and developing innovative applications in areas such as drug design, materials science and nanotechnology (De Paula, 2011).

1.6. GAS LAWS AND THE BEHAVIOR OF IDEAL GASES

The study of gases and their behavior is governed by a set of fundamental principles known as the gas laws. These laws describe the relationship between the properties of gases, such as pressure, volume, temperature and the number of particles. Understanding the gas laws is essential in various scientific and technological fields, as it enables scientists and engineers to predict and manipulate the behavior of gases. In this section, we will explore the gas laws and the behavior of ideal gases, which provide a simplified model for understanding gas properties.

The gas laws originated from the work of scientists such as Robert Boyle, Jacques Charles and Joseph Gay-Lussac, who conducted pioneering experiments in the 17th and 18th centuries. These scientists observed the relationship between different properties of gases and formulated laws that describe these relationships.

One of the fundamental gas laws is Boyle's law, which states that at constant temperature, the volume of a given amount of gas is inversely proportional to its pressure. Mathematically, this can be expressed as $P_1V_1 = P_2V_2$, where P_1 and V_1 are the initial pressure and volume and P_2 and V_2 are the final pressure and volume. Boyle's law suggests that as the volume of a gas decreases, its pressure increases and vice versa, as long as the temperature remains constant.

Charles's law, another important gas law, describes the relationship between the volume and temperature of a gas at constant pressure. Charles's law states that the volume of a given amount of gas is directly proportional to its absolute temperature. In equation form, this can be written as $V_1/T_1 = V_2/T_2$, where V_1 and T_1 are the initial volume and temperature and V_2 and T_2 are the final volume and temperature. Charles's law implies that as the temperature of a gas increases, its volume also increases, as long as the pressure remains constant.

Gay-Lussac's law, also known as the pressure-temperature law, states that the pressure of a given amount of gas is directly proportional to its absolute temperature at constant volume. Mathematically, this can be expressed as $P_1/T_1 = P_2/T_2$, where P_1 and T_1 are the initial pressure and temperature and P_2 and T_2 are the final pressure and temperature. Gay-Lussac's law suggests that as the temperature of a gas increases, its pressure also increases, assuming the volume remains constant (Adamson, 2012).

The three gas laws mentioned above can be combined into a single equation known as the ideal gas law. The ideal gas law relates the pressure, volume, temperature and number of particles of an ideal gas. It is expressed as $PV = nRT$, where P is the pressure, V is the volume, n is the number of moles of gas, R is the ideal gas constant and T is the temperature in Kelvin. The ideal gas law provides a general equation that describes the behavior of gases under various conditions (De Paula, 2011).

An ideal gas is a theoretical concept that assumes certain idealized conditions, including negligible molecular size, no intermolecular forces and elastic collisions. While no real gas perfectly fulfills these conditions, many gases behave reasonably close to the ideal gas behavior under normal conditions. The behavior of an ideal gas is governed by the kinetic theory of gases, which describes the motion and interactions of gas particles.

According to the kinetic theory of gases, gas particles are in constant random motion and possess kinetic energy. The average kinetic energy of gas particles is directly proportional to the temperature of the gas. Additionally, the pressure exerted by a gas is a result of the collisions between gas particles and the walls of the container. The greater the number of collisions and the higher the average kinetic energy of the particles, the higher the pressure of the gas.

The ideal gas law and the concept of ideal gases allow us to make predictions and perform calculations related to gas behavior. For example, the ideal gas law can be used to calculate the unknown value of one variable (pressure, volume, temperature, or number of moles) when the other three variables are known. This is particularly useful in solving problems related to gas properties, such as determining the volume of a gas at a given pressure and temperature or calculating the number of moles of gas based on its volume and temperature. The ideal gas law can also be used to understand the behavior of gases in various processes, such as isothermal, isobaric and isochoric processes. In an isothermal process, the temperature of the gas remains constant and the ideal gas law simplifies to $PV = \text{constant}$. In an isobaric process, the pressure of the gas remains constant and the ideal gas law simplifies to $V/T = \text{constant}$. In an isochoric process, the volume of the gas remains constant and the ideal gas law simplifies to $P/T = \text{constant}$. Deviation from ideal gas behavior can occur under certain conditions, particularly at high pressures or low temperatures. Real gases may exhibit deviations from the ideal gas law due to the presence of intermolecular forces and molecular size. These deviations can be accounted for by using modifications to the ideal gas law, such as the Van der Waals equation (Adamson, 2012).

1.7. STOICHIOMETRY AND CHEMICAL EQUATIONS

Stoichiometry is a fundamental concept in chemistry that deals with the quantitative relationships between reactants and products in a chemical reaction. It allows us to determine the amounts of substances involved in a reaction and predict the products formed. Chemical equations provide a symbolic representation of a chemical reaction, indicating the reactants and products involved. In this section, we will explore the principles of stoichiometry and how chemical equations are used to understand and analyze chemical reactions (De Paula, 2011).

Chemical equations provide a concise representation of a chemical reaction. They use chemical formulas and symbols to describe the reactants and products involved, as well as the stoichiometric relationship between them. A balanced chemical equation ensures that the number of atoms of each element is the same on both sides of the equation, reflecting the law of conservation of mass.

For example, the combustion of methane (CH_4) in oxygen (O_2) can be represented by the balanced chemical equation:

$$CH_4 + 2O_2 \rightarrow CO_2 + 2H_2O$$

No.	Stoichiometric reaction	Conversion (%)
1	Reduction: $2CrO_3 + Na_2S_2O_5 + 3H_2SO_4 \rightarrow Cr_2(SO_4)_3 + 2NaSO_4 + 3H_2O$	90.0
2	Neutralization 1: $Ca(OH)_2 + H_2SO_4 \rightarrow CaSO_4 + 2H_2O$	95.0
3	Precipitation: $3Ca(OH)_2 + Cr_2(SO_4)_3 \rightarrow 3CaSO_4 + 2Cr(OH)_3$	90.0
4	Neutralization 2: $2NaOH + H_2SO_4 \rightarrow Na_2SO_4 + 2H_2O$	95.0
5	Coagulation: $FeCl_3 + \dot{C}r(OH)_3 \rightarrow Fe(OH)_3 + CrCl_3$	98.0

Figure 1.4. Stoichiometry of chemical reactions.

Source: Pérez Sánchez, Amaury & González, Marisabel. (2022). Techno-economic evaluation of a chromic wastewater treatment plant proposal. Revista ION, 35, 33–48. Scientific Figure on ResearchGate. Available from: https://www.researchgate.net/figure/Stoichiometry-of-the-chemical-reactions-that-take-place-in-the-different-stages-of-the_tbl2_366077938 (accessed 22 September 2023)

This equation indicates that one molecule of methane reacts with two molecules of oxygen to produce one molecule of carbon dioxide and two molecules of water. The coefficients in front of the chemical formulas represent the stoichiometric ratios between the reactants and products (Figure 1.4).

Stoichiometry allows us to determine the quantities of substances involved in a chemical reaction. The stoichiometric coefficients in a balanced chemical equation provide the mole-to-mole ratio between the reactants and products. This ratio allows us to convert between the amounts of substances involved, using the concept of the mole, which represents a fixed number of particles (Avogadro's number, 6.022×10^{23}).

To perform stoichiometric calculations, we often start with the known amount of one substance (in moles or mass) and use the stoichiometric ratios to determine the amount of another substance involved. This process is commonly referred to as stoichiometric calculations or mole-to-mole conversions (De Paula, 2011).

For example, consider the reaction between hydrogen (H_2) and oxygen (O_2) to form water (H_2O):

$$2H_2 + O_2 \rightarrow 2H_2O$$

Suppose we have 4 moles of hydrogen gas and want to determine the amount of water produced. From the balanced equation, we can see that the stoichiometric ratio between hydrogen and water is 2:2, meaning that for every 2 moles of hydrogen gas, we obtain 2 moles of water.

Therefore, using the mole-to-mole ratio, we can calculate that 4 moles of hydrogen gas will produce 4 moles of water (Adamson, 2012). Stoichiometry calculations can also involve mass-to-mass conversions. In these calculations, we use the molar mass of a substance to convert between the mass of one substance and the mass of another substance involved in the reaction.

For example, let's consider the reaction between calcium carbonate ($CaCO_3$) and hydrochloric acid (HCl) to form calcium chloride ($CaCl_2$), carbon dioxide (CO_2) and water (H_2O):

$$CaCO_3 + 2HCl \rightarrow CaCl_2 + CO_2 + H_2O$$

Suppose we have 25 grams of calcium carbonate and want to determine the mass of carbon dioxide produced. First, we convert the mass of calcium carbonate to moles by dividing it by the molar mass of $CaCO_3$. Then, using the stoichiometric ratio between $CaCO_3$ and CO_2 (1:1), we can determine the moles of CO_2 produced. Finally, we convert the moles of CO_2 to grams by multiplying by the molar mass of CO_2.

Stoichiometry calculations are not limited to determining the amounts of reactants and products. They can also be used to predict the theoretical yield and percent yield of a reaction. The theoretical yield is the maximum amount of product that can be obtained based on stoichiometric calculations. The percent yield compares the actual yield (the amount of product obtained in a real experiment) to the theoretical yield and expresses it as a percentage (Adamson, 2012). Chemical equations also provide information about the limiting reactant in a reaction. The limiting reactant is the reactant that is completely consumed and determines the maximum amount of product that can be formed. The reactant in excess is the one that remains after the limiting reactant is completely consumed. Determining the limiting reactant is crucial for calculating the theoretical yield and understanding the efficiency of a reaction.

To determine the limiting reactant, we compare the moles or masses of the reactants involved in the reaction using stoichiometric calculations. The reactant that produces the least amount of product is the limiting reactant (De Paula, 2011).

For example, consider the reaction between 2 moles of hydrogen gas (H_2) and 1 mole of oxygen gas (O_2) to form water (H_2O):

$$2H_2 + O_2 \rightarrow 2H_2O$$

If we have 3 moles of hydrogen gas and 2 moles of oxygen gas, we can determine the limiting reactant by comparing the moles of water produced by each reactant. From the stoichiometric ratio, we know that 2 moles of hydrogen gas produce 2 moles of water, while 1 mole of oxygen gas produces 2 moles of water. Therefore, the hydrogen gas is the limiting reactant and the oxygen gas is in excess.

Understanding stoichiometry and chemical equations is essential for various applications in chemistry. It allows us to determine the amounts of reactants and products involved in a reaction, predict the theoretical yield and percent yield, identify the limiting reactant and perform stoichiometric calculations. Stoichiometry is crucial in fields such as analytical chemistry, pharmaceutical development, materials science and environmental science (Adamson, 2012).

1.8. ACID-BASE CONCEPTS AND PH CALCULATIONS

Acids and bases are fundamental concepts in chemistry that play a crucial role in various chemical reactions and biological processes. Understanding acid-base concepts and being able to calculate pH (a measure of acidity or alkalinity) are essential for comprehending chemical equilibrium, acid-base reactions and their implications in many areas of chemistry, biology and everyday life. In this section, we will explore acid-base concepts, pH calculations and their significance in chemical systems.

Acids and bases are defined based on their behavior in aqueous solutions. The most widely accepted definitions of acids and bases are the Arrhenius, Brønsted-Lowry and Lewis definitions. According to the Arrhenius definition, acids are substances that increase the concentration of hydrogen ions (H^+) when dissolved in water, while bases are substances that increase the concentration of hydroxide ions (OH^-). For example, when hydrochloric acid (HCl) dissolves in water, it donates H^+ ions, leading to an increase in the concentration of H^+ ions in the solution. On the other hand, when sodium hydroxide (NaOH) dissolves in water, it dissociates to produce OH^- ions, increasing the concentration of OH^- ions in the solution.

The Brønsted-Lowry definition focuses on the transfer of protons (H^+ ions) in acid-base reactions. According to this definition, an acid is a substance that donates a proton, while a base is a substance that accepts a proton. In this concept, acid-base reactions involve the transfer of protons between the acid and the base. For example, when hydrochloric acid (HCl) reacts with ammonia (NH_3), HCl donates a proton to NH_3, forming the ammonium ion (NH_4^+) and the chloride ion (Cl^-) (De Paula, 2011).

The Lewis definition of acids and bases is broader and encompasses a wider range of reactions. According to this definition, an acid is a substance that accepts an electron pair, while a base is a substance that donates an electron pair. In this concept, acid-base reactions involve the formation of coordinate covalent bonds, where the acid accepts a pair of electrons from the base. For example, in the reaction between boron trifluoride (BF_3) and ammonia (NH_3), BF_3 acts as an electron pair acceptor (Lewis acid), accepting a pair of electrons from the lone pair of electrons on NH_3 (Lewis base).

The concept of pH is a quantitative measure of the acidity or alkalinity of a solution. pH is defined as the negative logarithm (base 10) of the concentration of hydrogen ions in a solution. Mathematically, it can be expressed as $pH = -\log[H^+]$, where $[H^+]$ represents the concentration of hydrogen ions in moles per liter (Adamson, 2012).

The pH scale ranges from 0 to 14, with 7 considered neutral. A pH less than 7 indicates acidity, with lower pH values indicating stronger acidity. A pH greater than 7 indicates alkalinity or basicity, with higher pH values indicating stronger alkalinity. The pH scale is logarithmic, meaning that each unit change in pH represents a tenfold change in hydrogen ion concentration. For example, a solution with a pH of 4 is ten times more acidic than a solution with a pH of 5.

pH calculations can be performed using the concentration of hydrogen ions or hydroxide ions, depending on whether the solution is acidic or basic. In an acidic solution, the concentration of hydrogen ions is greater than that of hydroxide ions, while in a basic solution, the concentration of hydroxide ions is greater than that of hydrogen ions. To calculate the pH, we take the negative logarithm of the hydrogen ion concentration or the pOH (the negative logarithm of the hydroxide ion concentration) and convert it to pH using the relationship pH + pOH = 14.

For example, suppose we have a solution with a hydrogen ion concentration of 1 x 10^{-4} M. To calculate the pH, we take the negative logarithm of the hydrogen ion concentration:

$$pH = -\log(1 \times 10^{-4}) = 4$$

This indicates that the solution has a pH of 4, suggesting it is acidic (De Paula, 2011).

Acid-base reactions involve the transfer of protons from the acid to the base. The strength of an acid or base is determined by its ability to donate or accept protons. Strong acids, such as hydrochloric acid (HCl) and sulfuric acid (H_2SO_4), readily donate protons and completely dissociate in water. Strong bases, such as sodium hydroxide (NaOH) and potassium hydroxide (KOH), readily accept protons and dissociate to release hydroxide ions.

Weak acids and bases, on the other hand, only partially dissociate in water, resulting in an equilibrium between the undissociated species and the ions. The strength of an acid or base can be quantified using the concept of acid dissociation constant (Ka) or base dissociation constant (Kb). These constants represent the extent to which an acid or base dissociates in water. The larger the Ka or Kb value, the stronger the acid or base.

Acid-base reactions can be classified into different types, including neutralization reactions, acid-base titrations and buffer systems. Neutralization reactions occur when an acid and a base react to form a salt and water. For example, the reaction between hydrochloric acid (HCl) and sodium hydroxide (NaOH) forms sodium chloride (NaCl) and water (H_2O) (Adamson, 2012).

Acid-base titrations involve the gradual addition of a solution of known concentration (the titrant) to a solution of unknown concentration until the reaction between the acid and the base is complete. The point at which the stoichiometrically equivalent amounts of acid and base are reached is called the equivalence point. This allows us to determine the concentration of the unknown acid or base.

Buffer systems are important in maintaining the pH stability of solutions. A buffer is a solution that resists changes in pH when small amounts of acid or base are added. Buffers typically consist of a weak acid and its conjugate base (or a weak base and its conjugate acid). The weak acid can donate protons to neutralize added base, while the conjugate base can accept protons to neutralize added acid. Buffer systems are crucial in biological systems, as they help maintain the pH balance necessary for proper functioning (De Paula, 2011).

CHAPTER 2

THERMODYNAMICS

Contents

2.1. INTRODUCTION

Thermodynamics is a part of physics that studies how energy, heat and work are connected in different systems (Figure 2.1). It helps us understand and measure how energy changes, so we can study and predict how things behave. The laws of thermodynamics, along with concepts like enthalpy, entropy and Gibbs energy, are the basis of this area of study. Thermodynamics is an important part of science that helps us understand how energy changes and how materials behave. This helps us understand and foresee how different systems in engineering, chemistry and environmental science works. The three laws of thermodynamics are important rules that help us understand how energy works. They are connected to ideas like enthalpy, entropy and Gibbs energy. These concepts are used to study and use energy in different ways. Thermodynamics is important because it helps us understand the basic rules that control the movement and change of energy in different ways. By understanding these principles, we can learn about how things work in the world and use that knowledge to create new technologies and solve problems (Fermi, 2012). The First Law of Thermodynamics, also called the law of energy conservation, says that energy cannot be made or disappear. It can only change into different types. This principle helps us understand how energy moves and changes within a system. The first law helps us understand how energy moves and is used, from making electricity in power plants to running engines.

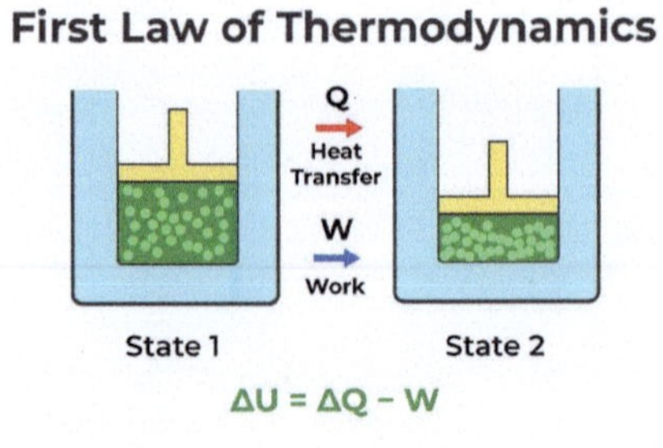

Figure 2.1. Basic principle of thermodynamics model.

Source: Marek, Laciak & Ján, Kačur & Karol, Kostúr. (2008). Simulation analysis for UCG with thermo dynamics model. ICCC'2008-proceedings of 9th International Carpathian Control Conference. 358–361. Scientific Figure on ResearchGate. Available from: https://www.researchgate.net/figure/Basic-principle-of-thermodynamics-model_fig1_306157637 (accessed on 22 September 2023)

The Second Law of Thermodynamics talks about entropy, which is a way to measure how messy a system is. This law says that some processes cannot be undone and that systems tend to become more disorganized over time. This helps us understand how well heat engines work and the limits of certain phenomena by limiting how energy can be transferred and converted. The Third Law of Thermodynamics studies what happens to things when they get really, really cold. It looks at how entropy, which is a measure of disorder, relates to temperature. This law helps us understand what materials are made of and how they act when they get really, really cold. When we understand the third law, we can understand things like superconductivity and superfluidity (Brewer, 2020).

Enthalpy is a term used in thermodynamics to describe the amount of heat in a system when the pressure is kept constant. This helps us learn how heat moves and how energy is involved in chemical reactions. Enthalpy is important in many areas, like making better heating and cooling systems and improving chemical processes. Entropy is a way to measure how messy or random something is in a system.

This idea is really important in thermodynamics and necessary for understanding the Second Law. By studying the changes in entropy, we can figure out which way processes happen naturally and understand what influences how a system behaves. Gibbs energy, also known as Gibbs free energy, combines the ideas of heat and disorder to figure out if a reaction will happen by itself and reach a balance. It helps us understand if chemical processes can work and find their balanced conditions. Gibbs energy helps us understand chemical systems better and helps us make chemical reactions and processes work better (Fermi, 2012).

Thermodynamics is useful in different areas besides understanding the basic principles and concepts. Engineering applications involve creating and designing things like heat engines, power plants, cooling systems and technologies that convert energy. In chemistry, thermodynamics helps us understand how chemical reactions happen, when things are balanced and if things will stay the same or change. In simple words, thermodynamics is very important in environmental science. It helps us understand renewable energy, climate change and how to store energy (Brewer, 2020).

2.2. FIRST LAW OF THERMODYNAMICS

The First Law of Thermodynamics is a very important rule in thermodynamics. It is also called the law of energy conservation. This says that energy cannot disappear or be made, it can only change shape or move between different things.

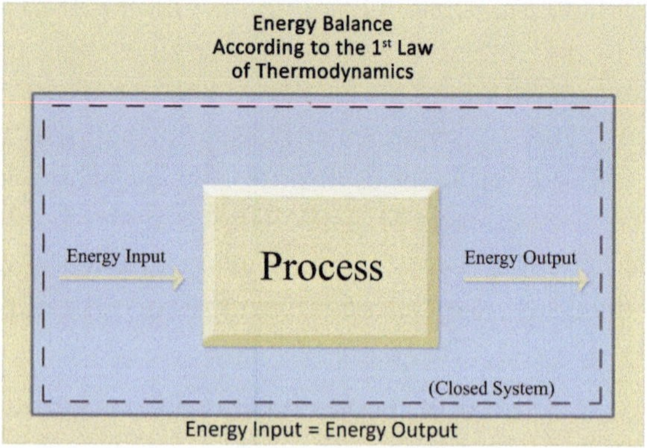

Figure 2.2. The first law of thermodynamics and its relationships with the expansion law of the universe and the Einstein field equations.

Source: Dezaki, Fatemeh & Mirza, Behrouz. (2014). The first law of thermodynamics as the origin of the expansion law of the universe. Scientific Figure on ResearchGate. Available from: https://www.researchgate.net/figure/The-first-law-of-thermodynamics-and-its-relationships-with-the-expansion-law-of-the_fig1_263163949 (accessed on 22 September 2023)

This law says that energy cannot be created or destroyed, just transferred. It helps us understand how energy moves in a system. The first law of thermodynamics says that energy cannot be created or destroyed in any process. It means that the amount of energy in a closed system doesn't change. This rule is true for any closed system, like a gas in a container or a chemical reaction.

To understand the first law, we need to talk about internal energy. Internal energy is the overall energy that a system has, which includes the energy from the movement and position of its particles. This refers to the tiny characteristics like how each tiny particle moves and interacts in the system. The first law says that the amount of energy a system has changes because of the heat added minus the work done. The change in internal energy (ΔU) is equal to the heat transferred (Q) minus the work done on or by the system (W). In simple words, Q means the heat going into the system and W means the work being done by the system. The first law says that when heat is added to a system, the internal energy goes up. But, when work is done by the system, the internal energy goes down (Fermi, 2012).

- Heat can move from one place to another in three different ways: through direct contact, through moving liquids or gases, or through waves in the air. Conduction is when heat is transferred through touching, while convection is when heat is transferred through the movement of liquids or gases. Radiation means sending out and taking in electromagnetic waves. When we talk about work done by the system, we mean the energy that gets transferred because of forces acting on it. It can come in different forms, like when gases get bigger or smaller, the use of electricity, or the work done by a machine.
- The first law of thermodynamics applies to everything, not just certain processes or systems. This information helps us understand how energy moves and is used in many different situations. For instance, in power plants, the first law is very important for studying how fuel energy gets turned into electricity. This helps measure how well energy is converted and helps us make better systems that use energy more efficiently (Brewer, 2020).
- Additionally, the first law of thermodynamics has important consequences in the study of heat engines. Heat engines change heat energy into movement power and the first law tells us how the amount of heat put into the engine affects how much work it can produce, as well as how efficient it is. The effectiveness of a heat engine is determined by: Efficiency is calculated by dividing the work output by the heat input and then multiplying by 100%. Understanding the first law helps engineers make better heat engine designs, make them work more efficiently and reduce energy waste.

In simple terms, the first law of thermodynamics, also called the law of energy conservation, means that energy cannot be created or destroyed in any process. This text emphasizes the importance of internal energy, the transfer of heat and the work performed by a system. By using the first law, we can study how energy moves, learn about how energy changes from one form to another and create better systems in many different areas like making electricity and building machines (Fermi, 2012).

2.3. SECOND LAW OF THERMODYNAMICS

The Second Law of Thermodynamics is a basic rule in thermodynamics that talks about the idea of entropy and how some processes cannot be undone. This helps us understand how energy moves and how systems tend to become messier or disordered. The second law is easier to grasp when we think about entropy. Entropy is a way to measure how messy or chaotic something is. This counts how many ways particles can be arranged in a specific state. Systems tend to change to states where things are more mixed up or chaotic (Brewer, 2020).

The second law of thermodynamics can be explained in different ways, but one common way is called the Clausius statement. It says that heat cannot naturally move from a colder object to a hotter object. This statement says that heat always moves from hot places to cold places and can't go the other way around. Another important idea that is related to the second law is the concept of a heat engine. A heat engine is a machine that changes heat energy into movement or mechanical work. The second law says that there are limits on how well heat engines can work. It means that no engine can be 100% efficient. This is called the Carnot statement of the second law.

The Carnot statement also talks about a reversible process. A reversible process is a special kind of process that can be undone without causing any lasting effects on the surroundings. In a reversible process, the system and its surroundings can go back to how they were at the beginning without causing any overall changes in entropy. The Carnot cycle is a set of steps that a heat engine can go through. These steps are reversible, which means they can go back and forth. This cycle shows us the highest possible efficiency that a heat engine can have when it operates between two different temperatures.

In real life, most processes cannot be undone completely. Irreversible actions cause the overall amount of disorder (entropy) in the system and its surroundings to go up. Examples of things that are irreversible include when things rub together and create heat, when heat moves from hot to cold areas and any process that happens quickly and not smoothly. The growth of disorder caused by irreversible processes is commonly called entropy generation. It measures how much energy cannot be used for useful tasks. This means that certain processes in the real world cannot be undone and cause disorder to increase over time. This idea shows that there is a difference between what has already happened and what will happen in the future (Fermi, 2012). The second law of thermodynamics has a big impact in many different areas. In engineering, it is important to understand the limits and effectiveness of energy conversion systems like power plants and heat engines. It helps engineers create better processes and use less energy. In chemistry, the second law predicts which way chemical reactions will naturally happen. It helps us figure out when reactions will happen and gives us information about how chemicals stay balanced and stable. In addition, the second law of thermodynamics has effects on subjects such as studying the universe and understanding the environment. It helps us learn about how the universe started and changed over time and it also helps us study complicated things like ecosystems and climate change (Brewer, 2020). In simple terms, the second law of thermodynamics tells us that energy always flows in certain ways and that systems naturally tend to become more disorganized over time. This sets limitations on how heat can move, how well engines can convert heat into work and how irreversible processes work. By knowing the second law, we can make energy conversion better, figure out if chemical reactions happen on their own and study complicated systems in science and engineering.

2.4. THIRD LAW OF THERMODYNAMICS

The Third Law of Thermodynamics is a basic rule in thermodynamics that helps us understand what happens to matter when it gets really, really cold. It also helps us see how entropy and temperature are related to each other. This statement says that when the temperature gets really, really cold, the entropy of a pure crystal-like substance gets as low as it can go (Fermi, 2012).

The third law of thermodynamics has a few things that happen because of it:

- Absolute zero is the coldest temperature ever. At this temperature, particles stop moving. The third law says that when a substance gets really, really cold, its entropy, which is its disorder, becomes zero. In simpler terms, when it's very cold, a perfect crystal would have no mess or chaos. Although it is not possible to reach absolute zero in theory, the third law of thermodynamics helps us understand how matter behaves at very low temperatures.
- Nernst's Heat Theorem is about the third law and it says that you can't ever reach absolute zero in a certain number of steps using a process that can go back and forth. This idea shows that absolute zero is impossible to reach and that there is always some leftover disorder, even at extremely cold temperatures (Brewer, 2020).
- The third law of thermodynamics affects how materials and systems behave when they are very cold. This helps us learn about things like superconductivity, where materials can have

no electrical resistance when it gets really cold. Furthermore, it helps us understand things like superfluidity and Bose-Einstein condensation, which happen at extremely cold temperatures and involve special characteristics of matter.

- The third law of thermodynamics affects how we calculate and measure entropy. By using absolute zero as a starting point, we can figure out how much entropy there is at different temperatures.
- In materials science and the study of phase transitions, the third law is really important. This thing helps us understand and predict how materials act when they get really cold. It helps us design and understand materials that have special properties.
- The third law has effects on computer simulations and modeling. It helps calculate entropy more accurately at low temperatures by providing a starting point for calculations.

The third law of thermodynamics talks about how things behave when they are really, really cold. But it also has implications for different areas of study. This research helps us learn more about materials science, how things change from one phase to another and the characteristics of substances. Moreover, it helps to calculate entropy and is useful for creating computer models and simulations. Basically, the third law of thermodynamics links entropy and temperature when the temperature gets very close to absolute zero. This shows information about how things act at really cold temperatures, is helpful for studying materials and assists in calculating entropy. When we understand the third law, it helps us understand the basic things about matter better. This understanding also helps us make progress in scientific and technological areas (Brewer, 2020).

2.5. ENTHALPY

Enthalpy is how we measure the heat content of a system when the pressure is constant. It adds up the energy inside a system with the energy gained or lost by the system through work to take into account changes in energy during processes.

Enthalpy means H. The heat energy of a system can be described using math. In simpler terms, H is equal to U plus PV. In simple terms, H is the enthalpy which measures the total energy of a system. U is the internal energy contained within the system. P refers to the pressure exerted on the system and V represents the volume of the system (Fermi, 2012).

Enthalpy is important because it helps us understand how heat is transferred and work is done in different processes. This helps us understand and measure the energy changes that happen during chemical reactions, changes in state and other processes related to temperature and energy. Enthalpy is often used to study how heat is transferred. The change in enthalpy (ΔH) tells us how much heat is either taken in or let out during a process. In reactions that release heat, the change in enthalpy is negative. On the other hand, in reactions that absorb heat from their surroundings, the change in enthalpy is positive ($\Delta H > 0$). The change in enthalpy is important for understanding how heat moves in chemical reactions and for designing processes that involve transferring heat, like in heaters or heat exchangers used in industries.

Enthalpy is important in chemical reactions and stoichiometry. The heat of reaction, also known as the enthalpy change, tells us how much energy is released or needed for a chemical reaction. This helps scientists figure out if reactions can happen and how much energy they need. It helps chemists make conditions better and predict how much heat is made or used. The measurement of heat created during a reaction is often done using a technique called calorimetry. This technique involves measuring any changes in temperature (Brewer, 2020).

Enthalpy is really important when things change from one phase to another, like when things get hot or cold, or when chemicals react with each other. When something converts from being solid to liquid or from liquid to gas, the energy required for this transformation is known as the heat of fusion or heat of vaporization. These changes in enthalpy show how much energy is needed or released when substances change from one form to another. This is important for things like learning materials and making sure they stay in good condition.

Enthalpy is a word that engineers use a lot when they want to design and make heating, ventilation and air conditioning systems better. Enthalpy calculations are useful for figuring out how much heat or cold is needed for a room. This helps to create HVAC systems that function effectively and give a lot of comfort without using excessive energy. In easy terms, enthalpy is a concept in thermodynamics that shows us the amount of heat in a system when the pressure remains the same. This calculates the overall energy in a system by adding or removing energy from the system through work. Enthalpy is helpful in transferring heat, chemical reactions, changes in various forms of matter and practical engineering uses. Understanding enthalpy helps us understand how energy changes, make things work better and create efficient systems in fields like chemistry, materials science and HVAC engineering (Fermi, 2012).

2.5.1. Enthalpy Change and Heat Transfer

The change in heat energy and the movement of heat are important ideas in the study of how things heat up and cool down and they are connected to each other. Enthalpy change (ΔH) shows how much heat is gained or lost in a process. Heat transfer is when thermal energy moves between systems. Enthalpy change means the difference in heat energy from the start to the end of something. It figures out how energy is spread with the environment when the pressure stays the same. We can measure the change in enthalpy using energy units like joules or calories. It can be either good or bad depending on if heat is absorbed or released (Brewer, 2020).

When heat is moved, the change in enthalpy shows how much heat energy was gained or lost. If the area around something is hot, then the change in heat is positive, which means the process takes in heat. In a process where the system absorbs heat, the system gets hotter and the surroundings get colder because heat is transferred into the system. On the other hand, if heat is given off to the surroundings, the change in enthalpy is negative, showing that it is a process that releases heat (exothermic). In an exothermic process, the system gets colder and the surroundings get hotter as heat moves out of the system. It is important to understand that enthalpy change includes both heat transfer and work done during a process. The work term is used to describe the exchange of mechanical energy, like when something expands or gets compressed. In certain real-life situations, like chemical reactions happening at a set pressure, the work done is usually not as important as the heat transferred and the change in enthalpy mostly shows the heat transfer. Heat transfer is when thermal energy moves between different systems. This happens in three main ways: through conduction, convection and radiation.

- Conduction is when heat moves from one thing to another by touching or going through a solid. It happens when there is a difference in temperature between two parts of a system or between two different systems touching each other. For instance, when you put a metal spoon in hot soup, the heat goes from the soup to the spoon through touching.
- Convection is when heat is transferred through the movement of liquids or gases. This means that it is the result or outcome when

both conduction (the transfer of heat through solid objects) and fluid motion (the movement of liquids or gases) happen together. Convection is when heat is transferred during boiling, when air rises from a hot surface and when a fan is used to make heat transfer better (Brewer, 2020).

- Radiation is the transferring of heat through energy waves. Radiation can happen without using any stuff, unlike conduction and convection. Radiation is when the Sun's warmth reaches Earth through waves or when something hot releases heat.

In simpler terms, enthalpy change refers to the amount of heat taken in or given off when something occurs and it also takes into account the work involved. Heat transfer is when heat moves from one object to another. It happens when we touch, move, or give off heat. It's important to understand the relationship between enthalpy change and heat transfer. This understanding helps us see how energy moves, improve things and make efficient systems in different thermodynamics situations (Fermi, 2012).

2.5.2. Enthalpy in Chemical Reactions

Enthalpy is important for studying and analyzing chemical reactions. This information helps us understand if a reaction releases or absorbs energy and allows us to estimate how much heat is involved. In a chemical reaction, the enthalpy change (ΔH) shows the difference in enthalpy between the starting materials and the resulting substances. It measures how much energy is exchanged with the surroundings. It can be positive or negative, which means it shows if heat is taken in or given off. An exothermic reaction is when there is a decrease in enthalpy ($\Delta H < 0$) and energy is released. In an exothermic reaction, heat is given off from the system to the surroundings. This heat release usually makes the surrounding area hotter. Some examples of exothermic reactions are when a fuel combines with oxygen to make heat and light, like in a fire.

On the other hand, an endothermic reaction is a reaction that absorbs heat ($\Delta H > 0$). In an endothermic reaction, the system takes in energy from the surroundings. This process usually results in the surroundings getting colder. Photosynthesis is an example of an endothermic reaction. It is when plants use energy from sunlight to turn carbon dioxide and water into glucose and oxygen. Enthalpy change can help us figure out how much heat is either released or taken in during a reaction. This can be accomplished by using a formula. In simple words, this equation means that the total heat change, denoted as ΔH, is equal to the product of the amount of substance involved, denoted as n and the heat capacity, denoted as q. Where heat (q) is either absorbed or released, enthalpy change (ΔH) represents the total energy change and the number of moles of the substance involved in the reaction (n) is considered. This equation helps us measure the amount of heat transferred during a chemical reaction and is important for creating and improving industrial processes (Brewer, 2020).

We can measure how much heat is given off or taken in during a chemical reaction by using a special tool called a calorimeter. Calorimetry is a way to measure how hot or cold something gets during a reaction. We use these measurements to figure out how much heat energy was given off or taken in during the reaction. Calorimetry helps us understand how energy changes in a reaction by accurately measuring the flow of heat. The changes in heat energy during chemical reactions can also be helpful in everyday life. These can be used to predict if reactions will work and occur naturally, assist in creating chemical processes and enhance the efficiency of reactions. Moreover, having knowledge about the changes in enthalpy is crucial in areas such as chemical engineering. This helps create and operate chemical reactors effectively by understanding how energy is transferred.

In simpler words, the change in enthalpy during chemical reactions helps us understand how energy is passed and heat is moved during the reaction. This helps us understand if a reaction is producing heat or absorbing heat and measures the quantity of heat involved. The enthalpy change measurement is helpful in figuring out if reactions can occur by themselves, discovering the optimal conditions for reactions and developing efficient chemical processes. It is vital to comprehend and apply this knowledge in areas like chemical engineering because conserving energy is incredibly important (Fermi, 2012).

2.6. ENTROPY

Entropy is a basic idea in thermodynamics that measures how much chaos or disorder there is in a system (Figure 2.3). It is very important for understanding how things work, both chemically and physically. It is also very important for the second law of thermodynamics. Entropy helps us understand which way energy moves, how natural processes happen and the restrictions on different events.

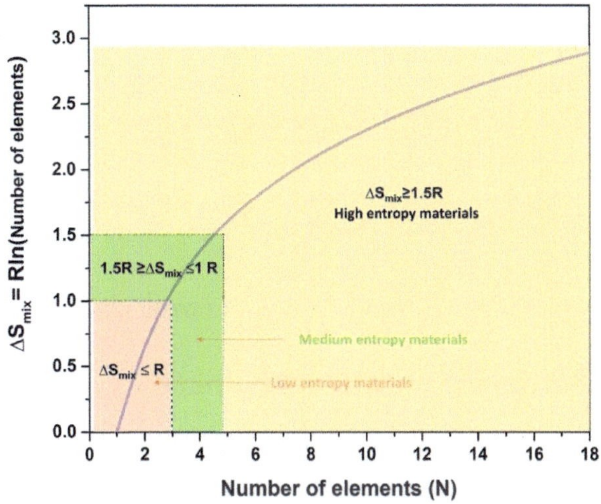

Figure 2.3. Variation of entropy with the number of elements and their subdivision in low, medium and high entropy materials.

Source: Kumar, Nirmal & Goel, Gaurav & Goel, Saurav. (2021). Emergence of machine learning in the development of high entropy alloy and their prospects in advanced engineering applications. Emergent Materials. Scientific Figure on ResearchGate. Available from: https://www.researchgate.net/figure/Variation-of-entropy-with-the-number-of-elements-and-their-subdivision-in-low-medium_fig1_352106240 (accessed on 22 September 2023)

The idea of entropy was developed as scientists tried to understand how energy works and the rules that govern how it changes. Entropy, a concept introduced by Rudolf Clausius in the 1800s, is now widely used in the study of thermodynamics. It helps us understand how different systems behave, from basic gases to complicated chemical reactions (Brewer, 2020).

Entropy and disorder are closely connected. When a system becomes messier, its disorder increases. It is important to understand that when entropy increases, it doesn't mean everything becomes chaotic. Instead, it means things become more random. Learning about entropy helps us predict how things will happen in science and chemistry. The second rule of thermodynamics talks about something called entropy. It says that when a system is isolated, its entropy will usually get bigger as time goes on. This law says certain processes cannot be undone and systems tend to become more disordered over time. It puts limits on how energy can be transferred and changed, showing the restrictions caused by entropy.

Entropy is useful in many different areas of science and engineering. In engineering, it is very important in creating and improving heat engines, refrigeration systems and devices that convert energy. In chemistry, entropy helps us understand how chemical reactions happen naturally and reach a balance. It helps us predict if a reaction will work and what products will be formed. Moreover, entropy is important in understanding the actions of complicated systems like ecosystems, biological happenings and even the whole universe. This has effects on topics like the study of how the universe started and changed and the examination of climate change (Fermi, 2012).

2.6.1. Entropy Change and Irreversibility

Entropy change (ΔS) measures the change in disorder between the starting and ending conditions of a system. It tells how messy or chaotic things are when something happens. Entropy change can go up or down, showing either more or less entropy. When things become messier or disorganized, the change in entropy is positive. This usually happens in processes where energy is scattered or spread out, which causes a more chaotic system. For instance, when a solid turn into a liquid, the tiny particles become less organized and the amount of disorderness, called entropy, increases (Brewer, 2020).

On the other hand, if things become more organized or predictable, it means there is a decrease in disorder, which is represented by a negative change in entropy ($\Delta S < 0$). This happens when things become more organized or when there is less spreading out of energy. For example, when a gas turns into a liquid, the particles become more orderly and the disorder (entropy) decreases.

Irreversibility means that some processes cannot be completely undone, even if everything else goes back to how it was before. Irreversible processes cause the total entropy of a system and its surroundings to go up. Energy is lost and becomes less useful during a process that cannot be reversed. This waste of energy causes the total disorder of the system to rise. Irreversible processes are ones that can't be undone and are caused by things like friction, heat transfer with temperature differences and any process that isn't smooth. Reversible processes are when things change very little and are used in theory to figure out the most efficient way a system can work.

Entropy change and irreversibility are strongly connected in thermodynamics. Irreversible processes always lead to more disorganization because energy is lost and not available anymore. The increase in randomness during a process that can't be reversed is a basic result of the second law of thermodynamics. Usually, irreversible processes lead to bigger, positive changes in entropy, which means there is more disorder and spreading out of energy. Reversible processes have very small changes in entropy because they involve tiny changes and do not result in an increase in disorder (Brewer, 2020).

It's important to understand that most real-world processes cannot be undone completely. It is usually difficult to make things perfectly reversible because of things like friction and heat transfer at different temperatures. However, reversible processes are a way to study and understand the limits of what is possible and to understand what happens when things are not reversible. So basically, entropy change and irreversibility are connected concepts in thermodynamics. Entropy change measures how disordered or random a system is, while irreversibility means processes that cannot be reversed exactly. Irreversible processes make the entropy go up, which means that energy is lost and cannot be used again. It is important to know the connection between the change in entropy and irreversibility to understand how a system behaves, predict which way processes will go and make systems work better (Fermi, 2012).

2.6.2. Entropy and the Second Law

Entropy and the Second Law of Thermodynamics are ideas that help us understand how things work and how energy moves in the universe. The second law tells us that things naturally move towards messier and more disordered states. This means that the disorder or randomness in an isolated system tends to increase as time goes on. It suggests that natural events tend to become more chaotic over time. Entropy is a way to describe how messy or random something is.

Systems with high entropy are much disorganized, while systems with low entropy are more organized and structured.

The second law means that things tend to become messier and more disorganized over time. Additionally, entropy is important in understanding how energy moves and changes. When energy moves from one place to another, like when heat is transferred or work is done, there is usually a change in the amount of disorder in the system, which is called entropy. For instance, when heat moves from something hot to something cold, disorder increases as energy spreads out more. Entropy is connected to things that can't be undone.

Irreversible processes cause the total entropy of the system and its surroundings to go up, while reversible processes have only small changes in entropy. The second law of thermodynamics controls how energy is used and transformed. It tells us the best efficiency we can achieve in different processes. Moreover, entropy and the second law are related to information theory. In information theory, entropy is used to measure the amount of information in a system. In simple terms, entropy and the second law of thermodynamics help us understand how processes work, why things tend to become more chaotic and the restrictions on changing energy (Brewer, 2020).

2.7. GIBBS ENERGY

Gibbs energy, or Gibbs free energy, is a scientific term used to describe a property that helps us understand how chemical reactions happen and reach a balance (Figure 2.4). This concept, named after scientist Josiah Willard Gibbs, helps us understand how much energy can be used to do something useful during a process. The Gibbs energy is a useful tool for figuring out if reactions can happen and which way they will go and for understanding how to make things balance. Gibbs energy is a measure that combines the ideas of heat energy and disorder in a substance into one value called G. In simple terms, this equation is written as G = H – TS. The letter T stands for the temperature measured in Kelvin. This equation shows how enthalpy, entropy and Gibbs energy are related and how they affect chemical reactions (Fermi, 2012).

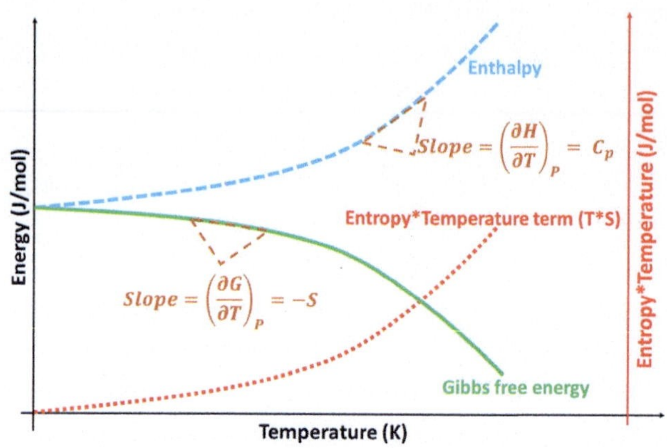

Figure 2.4. Diagram showing the variation of Gibbs free energy, enthalpy and entropy with temperature for a crystalline solid material.

Source: Hadjittofis, Eftychios & Isbell, Mark & Karde, Vikram & Varghese, Sophia & Ghoroi, Chinmay & Heng, Jerry. (2018). Influences of Crystal Anisotropy in Pharmaceutical Process Development. Pharmaceutical Research. Scientific Figure on ResearchGate. Available from: https://www.researchgate.net/figure/Diagram-showing-the-variation-of-Gibbs-free-energy-enthalpy-and-entropy-with_fig1_323861909 (accessed on 22 September 2023)

The Gibbs energy is used to determine if a reaction will happen on its own or not. According to the second law of thermodynamics, a spontaneous process happens by itself and makes everything more disordered. If the Gibbs energy of a reaction decreases (ΔG is negative), then the reaction happens naturally. A negative ΔG means that the reaction can go forward, releasing energy and making the system more disorganized. If ΔG is positive, it means that the reaction cannot happen on its own and it will happen in the opposite direction. Equilibrium is when ΔG is zero, meaning the system is balanced and the forward and reverse reactions happen at the same speed. When a system is at equilibrium, the Gibbs energy tells us how much work we can get from that system.

Gibbs energy helps us see if reactions will happen on their own and reach a balance. It also tells us what conditions are needed for good results. When we change the equation $\Delta G = \Delta H - T\Delta S$, we see that temperature, enthalpy and entropy all affect the value of ΔG. Changes in temperature, pressure, or the number of reactants can change the value of ΔG and also change how a reaction goes and how much of it happens. Gibbs energy is used in many areas like chemical engineering, materials science and biology, not just in chemical reactions. This helps with designing and improving chemical processes, understanding how materials change from one phase to another and learning about reactions in living things and how enzymes work (Brewer, 2020).

2.7.1. Gibbs Free Energy Change

The Gibbs free energy change, called ΔG, is a measure of the available energy to do useful work in a system when it goes through a chemical or physical change at a set temperature and pressure. It is an important factor in figuring out if a reaction or process can happen on its own and if it is possible. The Gibbs free energy change is a way to measure how a reaction or process will happen by considering both the energy released or absorbed and the randomness or disorder of the system. It is calculated using the equation $\Delta G = \Delta H - T\Delta S$, where ΔH is the change in energy, ΔS is the change in disorder and T is the temperature in Kelvin (Fermi, 2012). The ΔG sign shows if a reaction can happen on its own. If ΔG is negative ($\Delta G < 0$), it means the reaction happens without outside energy and the forward reaction is likely to occur. In a spontaneous process, the system gives out energy, usually as heat and the overall disorder of the system increases. However, if ΔG is positive ($\Delta G > 0$), the reaction does not happen naturally and the opposite reaction is preferred. When ΔG is zero, it means that the system is balanced and there is no overall change happening.

Relationship with Equilibrium: The Gibbs free energy change tells us what conditions are needed for a reaction to reach a balanced state. At equilibrium, the value of ΔG is zero, which means that the rates of the forward and reverse reactions are the same. If ΔG is negative, the reaction will go on its own in the forward direction until it reaches a balance. On the other hand, if ΔG is positive, the reaction will go in the opposite direction until it reaches a balance.

Importance of ΔH and ΔS: Understanding the changes in energy during a process involves looking at the contributions of enthalpy (ΔH) and entropy (ΔS) to ΔG. H represents the heat that is either taken in or given off during a reaction. Positive H means that energy is taken in (endothermic), while negative H means that energy is given off (exothermic). ΔS represents how much the system's disorder or randomness has changed. An increase in entropy means things become more disordered and this is beneficial because it promotes spontaneity and disperses energy (Brewer, 2020). The temperature of a substance impacts the total amount of ΔG. Temperature changes can affect the likelihood of a reaction occurring by itself. When it's very cold, the amount of heat change is the most important factor in the equation. Reactions that give off heat and have a negative change in energy (ΔH) are more likely to occur by themselves. When it is very hot, the change in disorder (ΔS) becomes the most important factor and reactions that make things untidier or disorganized are more likely to occur naturally (Fermi, 2012).

Using chemistry and engineering to solve real-world problems and improve industries. The change in Gibbs free energy is often used in chemistry and engineering. This helps determine if chemical reactions will be successful and in which direction they will proceed. This knowledge is useful for planning and improving manufacturing processes in industries.

In simple terms, the Gibbs free energy change (ΔG) tells us how much usable energy is in a system. It tells us if reactions will happen naturally and which way they will go. It also helps us understand when reactions are balanced. ΔG is a measure that takes into account the energy (ΔH) and disorder (ΔS) of a system. It also depends on the temperature (T). Knowing the value of ΔG is very important for figuring out if a reaction can happen, creating efficient processes and making the most of energy in different scientific and engineering uses (Brewer, 2020).

2.8. PRACTICAL APPLICATIONS OF THERMODYNAMICS

Thermodynamics is a part of physics that is used in many different areas of science, engineering and industry. Thermodynamics is important for understanding and improving how things work in our everyday lives. It helps us with things like making energy more efficient, designing better materials and improving how things are made.

Thermodynamics has an important role in producing and converting energy in real-life situations. Thermodynamic principles help in designing and operating power plants, engines and renewable energy systems. Engineers use thermodynamic ideas to make sure energy is used efficiently, reduce wasted energy and make processes like burning, transferring heat and storing energy work as well as possible. Thermodynamics is very important in chemical engineering because it helps with creating and improving chemical processes. Engineers use the laws of heat and energy to find the best conditions for chemical reactions, such as the right temperature, pressure and amounts of reactants. This makes sure that we can make products we want without using too much energy or creating too much waste.

Materials science is a subject that studies materials and it uses thermodynamics to help understand and predict how materials will behave. Thermodynamic concepts help us understand how materials change and how they are structured. They give us information about stability, balance between different phases and how things are formed at a microscopic level. This information helps scientists and engineers create new materials with specific qualities and improve how materials are made (Fermi, 2012).

Thermodynamics is also used in environmental science and sustainability. This tool helps understand and reduce the negative effects on the environment caused by industrial processes. It also helps create better and more environmentally-friendly technologies. Thermodynamic models are used to understand and study pollutants, measure greenhouse gas emissions and improve how energy is used in buildings and transportation. In biophysics and bioengineering, thermodynamics helps explain how biological processes work, like how enzymes work, how proteins fold and how cells produce energy. This helps us understand and create models of how energy moves and changes in living things (Brewer, 2020).

In addition, thermodynamics is used in aerospace engineering to help design engines, rockets and space systems. The principles of thermodynamics are used to make propulsion systems, fuel efficiency and thermal management better in these high-performance applications.

Basically, thermodynamics is used in many different areas. This supports making energy and changing it, helps with making chemicals and materials, helps with saving the environment, helps understand living things and affects making things for air travel. By using thermodynamic principles, scientists and engineers can make systems work better, use less energy and last longer. This helps create new technology and solve real-life problems (Fermi, 2012).

CHAPTER 3

Chemical Equilibria

Contents

3.1. INTRODUCTION

Chemical equilibrium is the state that exists when there is no net change over time in the concentration of the reactants and products involved in a chemical process. Another name for chemical equilibrium is "steady state reaction." This does not necessarily suggest that the chemical process has halted, but rather that there is now a balance between the absorption and production of chemicals. Although the ratio of reactant to product amounts has become stable, they are nearly never equal.

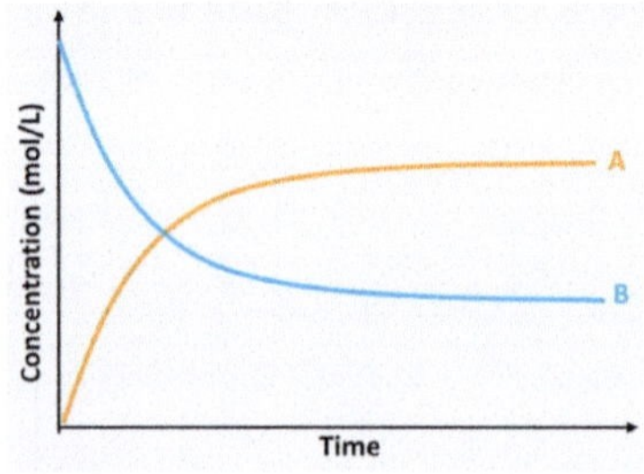

Figure 3.1. Chemical equilibrium. A: products; B: reactants.

Source: Furlan, Jacqueline & Siqueira, Adriano & Romão, Estaner. (2023). Simplified Model for Concentration Analysis of Catalytic Conversion of Carbon Gas into Hydrogen. Sustainability. Scientific Figure on ResearchGate. Available from: https://www.researchgate.net/figure/Chemical-Equilibrium-A-reactantes-B-products_fig2_366971371 (accessed on 22 September 2023)

There might be a lot more reactant or a lot more product. When the chemical reaction continues to take place but a proportion of the products and reactants stay constant, dynamic equilibrium develops (Figure 3.1). One example of chemical equilibrium is this (Storey, 2011).

The concentration of the products and reactants can be used to represent the equilibrium expression for a chemical process. Since the concentrations of liquids and solids do not vary, the equilibrium expression only takes into account chemical species in the aqueous and gaseous phases.

$jA + kB \rightarrow lC + mD$

The equation for the equilibrium is

$K = ([C]l[D]m) / ([A]j[B]k)$

The equilibrium constant is K.

The molecular concentrations of A, B, C, D etc. are [A], [B], [C], [D], etc. In a balanced chemical equation, coefficients like j, k, l, m, etc (Morgan, 2012). When there is chemical equilibrium, forward and reverse reactions are equal. Compounds C and D are formed when A and B combine. The reaction is A plus B equals C plus D. Because A and B interact to produce C and D, C and D may react to generate A and B. The solution is A+B+C+D. These two equations can be used to advance and reverse chemical reactions. In contrast to the reverse reaction, which transforms products into reactants ($C + D \rightarrow A + B$.), the forward reaction transforms reactants into products ($A + B \rightarrow C + D$).

Once the forward response has begun, numerous products emerge. Consequently, the opposite reaction begins at zero. When C and D arrive and their concentrations increase, the forward reaction slows down and the reverse reaction speeds up. This delay is due to a decrease in the concentration of A and B. Since the reaction continues to consume A and B to produce C and D, the rates of A and B reacting to produce C and D and C and D reacting to produce A and B are equal. In motion chemical equilibrium. They are dynamic because both forward and reverse reactions occur at the same concentrations of reactants and products. Chemical dynamics is comparable to a game of tug-of-war between you and a friend in which neither of you desires to prevail. There are three fundamental chemical equilibrium types (Storey, 2011).

The existence of chemical equilibrium in a closed system

- Reversible chemical equilibrium
- Chemical equilibrium at limited velocities

In a chemically stable closed system, the concentrations of reactants and products remain constant. Reversible reactions are equilibrium-forming reactions. In a slow reaction, the forward and reverse reactions are not equilibrium processes.

3.1.1. Conditions for Equilibrium

Chemical equilibrium is dynamic. After equilibrium, forward and reverse reactions continue. In equilibrium, reactants and products have the same relative concentrations and the same reaction rates. System equilibrium is contingent upon three conditions:

- The system net force is zero
- There is no system torque.
- The system operates at a constant rate.

- No chemicals may enter or leave the closed system.
- Dynamic equilibrium. Even though we can't see them, forward and backward responses are happening.
- Reactions must be equal in both directions.

Reactants and products might differ. After equilibrium, reactants and products will remain constant (Storey, 2011)

3.1.2. Factors That Affect Chemical Equilibrium

First, think of something that doesn't interfere with equilibrium: pure substances. When an equilibrium exists, a pure liquid or solid is thought to have an equilibrium constant of 1 and is not included in the equilibrium constant. For instance, pure water is thought to have an activity of 1, except for very concentrated solutions. Another illustration is solid carbon, which may be created when two molecules of carbon monoxide combine to produce carbon dioxide and carbon. Among the elements that do impact equilibrium are:

- Equilibrium is impacted by the addition of a reactant, product, or change in concentration. A chemical equation's equilibrium can be moved to the right, where more product develops, by adding a reactant. As additional reactant develops, adding product might shift the equilibrium to the left (Morgan, 2012).
- Equilibrium changes when the temperature changes. Chemical equilibrium is constantly shifted toward the endothermic reaction as the temperature rises. Equilibrium is constantly shifted in favor of the exothermic process as temperature decreases.
- Changes in pressure have an impact on balance. For instance, when the volume of a gas system is reduced, the pressure is increased, which raises the reactant and product concentrations. The whole reaction will result in a decrease in gas molecule concentration.

By adding stress to the system, the shift in equilibrium results may be predicted using Le Chatelier's concept. According to Le Chatelier's principle, every alteration to a system that is already in equilibrium will result in a predicted shift in equilibrium.

For instance, increasing the quantity of heat in a system favors the endothermic reaction's direction since it will cause the heat to be dissipated. A careful balancing act resembles chemical equilibrium in a chemical process. It occurs when the quantities of the initial constituents (reactants) and the final compounds (products) remain constant across time. When both the forward and backward responses move at the same rate, a form of steady state is reached (Storey, 2011).

Equilibrium is not a static condition, though, when nothing is happening. The reactions are more like a dynamic dance between the reactants and products, continuing at the same rate for both. To characterize this situation, scientists employ a concept known as the equilibrium constant (K). It is a ratio of the product concentrations to the reactant concentrations, with each concentration being increased to the power of the molecular weight of the corresponding reactant. This constant reveals the location of the equilibrium. If K is significantly more than 1, there are likely many more products and the reaction is strongly favoring that outcome. If K is much below 1, there are mostly reactants present. K indicates that both sides of the reaction are equally represented when it is near 1 (Morgan, 2012).

Various things may have an impact on the equilibrium position. The equilibrium can be shifted in one direction or another by altering the temperature, pressure (for processes involving gases), or the addition or removal of reactants or products. According to Le Chatelier's principle, if we upset an equilibrium, the system will attempt to balance itself out in response.

Chemical equilibrium is crucial to many chemistry-related processes. It aids in our comprehension of solubility, complex forms, gas-phase reactions and acid-base interactions. Because it enables scientists and engineers to design processes that are effective and limit waste, it is particularly crucial in industrial processes and environmental chemistry. In other words, chemical equilibrium is a dance between reactants and products where they strike a balance and maintain a constant quantity. It's an important idea that helps us comprehend and regulate chemical processes in a variety of contexts (Storey, 2011).

3.2. TYPES OF CHEMICAL EQUILIBRIUM

There are two types of chemical equilibrium (Figure 3.2):

- homogeneous equilibrium; and
- heterogeneous equilibrium.

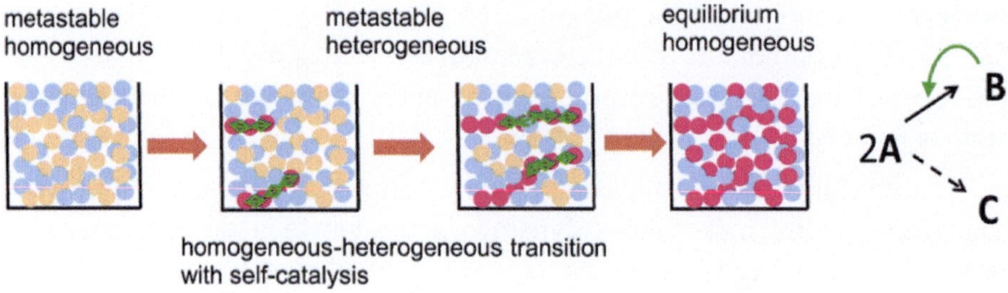

Figure 3.2. Spatially homogeneous–heterogeneous transition in a self-catalytic chemical reaction in dilute solution.

Source: Sawato, Tsukasa & Saito, Nozomi & Yamaguchi, Masahiko. (2019). Chemical Systems Involving Two Competitive Self-Catalytic Reactions. ACS Omega. Scientific Figure on ResearchGate. Available from: https://www.researchgate.net/figure/Spatially-homogeneous-heterogeneous-transition-in-a-self-catalytic-chemical-reaction-in_fig16_332000067 (accessed on 22 September 2023)

3.2.1. Homogeneous Chemical Equilibrium

Chemical equilibrium reactants and products are in the same phase in this process. It has two types:

(i) Reactions with equal product and reactant molecules.

- $H_2(g) + I_2(g) \rightleftharpoons 2HI(g)$
- $N_2(g) + O_2(g) \rightleftharpoons 2NO(g)$

(ii) Reactions where product molecules exceed reactant molecules.

- $2SO_2(g) + O_2(g) \rightleftharpoons 2SO_3(g)$
- $COCl_2(g) \rightleftharpoons CO(g) + Cl_2(g)$

Homogeneous chemical equilibrium balances reactant and product concentrations. The equilibrium constant is the ratio of reactants and products in a solution without any change in concentration or activity. In homogeneous chemical equilibrium, all chemical system constituents are in the same phase (gas, liquid, solid, or solution). Changing temperature or pressure creates this balance. The system is thermodynamically stable since its elements converge toward equilibrium. Homogeneous chemical equilibrium is the state of a chemical mixture in which all reactants attain an equilibrium concentration. If a process produces 4 moles of water and 2 moles of oxygen gas (Storey, 2011). Aqueous solutions have homogenous chemical equilibrium. Conditions affect these solutions. Water and air reacting in balance is not spontaneous. Thermodynamic equilibrium requires homogeneous chemical equilibrium, although chemical equilibrium does not. It optimizes reaction yield (Morgan, 2012).

Homogeneous chemical equilibrium factors

Chemistry's most explored topic is homogeneous chemical equilibrium. Pressure, volume, solubility and temperature affect homogenous chemical equilibrium.

- Temperature: As molecules loosen and have more space to interact, chemical reactions speed up. Because molecules become more closely linked and have less space to interact, chemical reaction rates fall as temperature decreases.
- Pressure: The equilibrium constant for a homogeneous chemical equilibrium is the ratio of reactant and product concentrations in equilibrium and is independent of pressure. The equilibrium gas can only be in one place since pressure changes the total gas in the system. Pressure decreases reactants and increases products and vice versa. Homogeneous equilibrium always has maximum pressure. Pressure homogenizes gas equilibrium.
- Volume: The volume dependence of homogeneous chemical equilibrium is generally stated as Q = kVn, where Q is the reaction heat, k is a constant and n is the number of moles of the reaction product (or solute). In a homogeneous system of a chemical element of known composition, the fraction of atoms of a given component in a compound remains constant while the system volume changes. As system volume varies, the ratio of chemical element concentrations in a compound will not change.
- Concentration: Chemical concentration shifts equilibrium to the right. Chemical concentration increases ion concentration. Homogeneous or single-phase aqueous solutions are dissolved acids in water. Many biological and inorganic systems contain aqueous solutions. Homogeneous solutions without dissociated ions are not solutions. Hydrogen-oxygen gas is in balance. If one gas is removed, the reaction continues until equilibrium. No third gas forms from the two gases. No reaction occurs until equilibrium is attained with equal amounts of water (Storey, 2011).

Homogeneous chemical equilibrium explains how chemical bonds develop. For instance, solid atoms must be in chemical equilibrium like liquid atoms (Morgan, 2012).

3.2.2. Heterogeneous Chemical Equilibrium

This reaction comprises reactants and chemical equilibrium products in various stages. Examples follow:

CO2(g) + C(s) ⇌ 2CO(g)

CaCO3(s) ⇌ CaO(s) + CO2(g)

3.3. FACTORS AFFECTING CHEMICAL EQUILIBRIUM

Le-Chatelier's principle states that if equilibrium circumstances change, the system will counteract or lessen the total transformation. This applies to chemical and physical balance. Temperature, pressure and system concentration impact equilibrium.

- **Change in pressure**

  Volume changes pressure. Pressure changes impact gaseous reactions because the number of reactants and products changes. According to Le Chatelier's principle, pressure changes in liquids and solids may be neglected in heterogeneous chemical equilibrium since volume is independent of pressure.

- **Change in temperature**

  Temperature affects chemical equilibrium according to Le-Chatelier's Principle and the reaction's ΔH. Exothermic reaction equilibrium constants decrease with temperature. Temperature raises the equilibrium constant for endothermic reactions. Temperature affects reaction rate and equilibrium constant. According to Le Chatelier's principle, exothermic reactions shift toward reactants as temperature rises. As temperature rises, endothermic processes move toward the product. The van 't Hoff equation explains how temperature affects equilibrium constants. K decreases during exothermic processes (ΔH is negative) but rises during endothermic processes (ΔH is positive). This appears to be a method for determining the standard molar enthalpy of a reaction by observing the temperature dependence of K. Due to error propagation, this method is insecure. In heterogeneous equilibrium, reactants and products exist in separate phases, whereas inhomogeneous equilibrium, they exist in the same phase. The Haber-Bosch process (right) uses equilibrium and adsorption phases to produce ammonia. Chemical oceanography logs for two-phase seawater D distribution coefficient: required for lipophilic medications. Chromatography, liquid-liquid extraction, ion exchange soluble material oxygenation and hemoglobin release are all examples of separation techniques. The components of acid-base equilibrium include buffers, indicators, hydrolysis and acid-base homeostasis. In the processes of segregation, chelation, MRI contrast agents and Schlenk equilibrium, metal-ligand complexation can be observed. Formation of an adduct requires molecular recognition, supramolecular chemistry and dinitrogen tetroxide. Some oscillatory systems achieve equilibrium when dampened oscillations reach equilibrium. The Nernst equation is used to plot the electrode potential versus redox concentration in electrochemistry (Storey, 2011).

When molecules on either side of an equilibrium endure irreversible secondary reactions, the Curtin-Hammett principle determines the final product ratio. These applications utilize constants associated with stability, formation, binding, affinity, association and dissociation. The concentration units used to calculate the value of a constant in biochemistry are binding constant units.

- **Effect of a Catalyst**

  No catalyst affects chemical equilibrium. It just accelerates reactions. Catalyzed and non-catalyzed reactions have the same reactants and products at equilibrium. A catalyst only helps when reactants to products transfer via a lower-energy state (Morgan, 2012).

- **Inert Gas Addition**

  Inert gases like argon introduced to a fixed volume do not react, hence the equilibrium remains unchanged. Gas supplied as a reactant or product changes the reaction quotient. Chemical reactions involve forward and reverse reactions. Reactants and products have distinct compositions. After a while, the forward and backward reflexes may equalize. After this, the reverse reaction will create the same amount of reactants, maintaining reactant and product concentrations. Thus, reactants and products will be chemically balanced. Chemical equilibrium occurs when the forward and backward reaction rates are equivalent and reactant and product concentrations are equal. Chemical equilibrium occurs when reaction rates are equal and reactants.

At the start of a reversible chemical process, any reaction can go faster forward or backward, but none can continue. The speeds of reactions equal each other over time if the reversible chemical reaction happens in a closed system and other parameters are constant. Thus, equilibrium always occurs when a chemical process begins. Equilibrium occurs when forward and backward reactions are equal. When this happens, the system in chemical equilibrium maintains its attributes. The system maintains equilibrium throughout time after reaching equilibrium. The system will fight to negate external constraints like changes in temperature, concentration, or pressure to preserve equilibrium. Understanding chemical equilibrium maintenance is crucial. To preserve chemical equilibrium in reversible processes. The shift in equilibrium position is the major reason reversible systems retain equilibrium. When disturbed, reversible chemical reactions alter the equilibrium position to preserve equilibrium. Le Chatelier's Principle openly defends equilibrium shift. Le Chatelier's Principle states that a system in equilibrium will move to counteract an external limitation such as a change in temperature, pressure, or concentration (Storey, 2011).

3.4. ADDITION OF REACTANTS OR PRODUCTS

A reaction system at equilibrium has Qr = Keq and eq = The equilibrium constant and reaction quotient differ if component activities change. Keq – Qr and then Increasing reagent I activity reaction quotient drops then and the reaction will go right (producing more products) product j activity rises and the reaction will move left, producing fewer products activities and equilibrium constants are dimensionless numbers (Morgan, 2012).

3.5. TREATMENT OF ACTIVITY

Activity coefficients may be calculated. Debye–Hückel equations, Davies equation extensions, specific ion interaction theories and Pitzer equations can be utilized to solve problems. Using the concentration quotient instead of the thermodynamic equilibrium constant assumes that is a constant. Instead of the more exact concentration quotient, the equilibrium constant is commonly used. Partial pressure replaces concentration and the fugacity coefficient replaces the activity coefficient for gas phase processes. Industrial ammonia production requires fugacity coefficients. Fugacity is the thermodynamic composition of a real gas. A species' actual gas phase chemical potential is hence the equilibrium constant formulation applies to solution and gas phases.

3.6. CONCENTRATION QUOTIENTS

In an aqueous solution, equilibrium constants are commonly measured with an "inert" electrolyte like sodium nitrate, $NaNO_3$, or potassium perchlorate, $KClO_4$. Solution ionic strength = where ci and zi are the concentration and ionic charge of ion type I and the total is over all N charged species in solution. When the dissolved salt concentration is substantially greater than the analytical concentrations of the reagents, the dissolved salt ions dictate the ionic strength, which is virtually constant. Species activity coefficients are independent of concentration since they depend on ionic strength. Thus, assuming is constant is justified. Concentration quotients are simple multiples of equilibrium constants. Ionic strength affects Kc. If measured at many ionic strengths, the value can be extrapolated to zero. This concentration quotient is paradoxically a thermodynamic equilibrium constant. Adjust a published equilibrium constant value before utilizing it in different ionic strength conditions (Storey, 2011).

3.6.1. Metastable Mixtures

Though not in equilibrium, a mixture may look stable. The kinetic barrier to SO3 production makes a combination of SO_2 and O_2 metastable.

$$2\ SO_2 + O_2 \rightleftharpoons 2\ SO_3$$

As with the contact process, adding a catalyst overcomes the barrier, but it does not change equilibrium concentrations. Under normal conditions, carbon dioxide and water create bicarbonate slowly.

$$CO_2 + 2\ H_2O \rightleftharpoons HCO_3^- + H_3O^+$$

However carbonic anhydrase speeds up the reaction (Morgan, 2012).

3.6.2. Pure Substances

Because pure liquids and solids have numerical values of one, their activities do not show in the equilibrium constant. The equilibrium constant for a dilute acetic acid solution in water is

$CH_3COOH + H_2O \rightleftharpoons CH_3COO^- + H_3O^+$

Water is a "pure" liquid with an activity of one in most solutions. Thus, the equilibrium constant expression is typical. Water self-ionizes.

$2\ H_2O \rightleftharpoons H_3O^+ + OH^-$

Since the proton's solvation state is constant in dilute solutions, the equilibrium concentrations are unaffected by writing $[H^+]$ for the hydronium ion concentration. Ionic strength and temperature affect Kw. H^+ and OH^- concentrations are related. Most equilibrium constant formulas incorporate hydroxide ion substitute $[OH^-]$ with $Kw\ [H^+]\ -1$. Pure solids with one activity do not appear in the equilibrium constant statement. The Boudouard reaction:

$2\ CO \rightleftharpoons CO_2 + C$

3.7. MULTIPLE EQUILIBRIA

Take dibasic acid H_2A. Water dissolves H_2A, HA^- and A^{2-} Two proton-releasing processes make up this equilibrium. Stepwise equilibrium constants include *K*1 and *K2*. Stepwise constants provide the equilibrium constant, βD. Dissociation products on the right side of the equilibrium formula make these constants dissociation constants. Many systems favor association constants. Association constants β1 and β2. Log β1 = pK2 and log β2 = pK2 + pK1

- **Composition of a mixture**

If the only equilibrium of a mixture is the formation of a 1:1 adduct, its composition can be approximated using a variety of methods. Refer to the ICE table for a well-known method of estimating the pH of mild acid solutions. There are three ways to determine a mixture's equilibrium composition (Storey, 2011).

- Changing equilibrium constants until the requisite concentrations are defined in terms of observable chemical potentials and initial conditions is the simplest method.
- Reduce the Gibbs energy of the system. Consider the weight. As demonstrated by mass balancing equations, according to the principle of mass conservation, the total concentration of each reactant must remain constant.

3.8. FORMULAS FOR MASS EQUILIBRIUM

Frequently, calculations are challenging. The proton H^+ and the conjugate base A^{2-}, interact with water containing H_2A. In the mass-balance calculations presented below, reactant A could be a base such as 1,2-diaminoethane. Containing a substantial amount of TA species A infrequently do these calculations account for ionic charges. The equilibrium constants and total concentrations determine two equations in two unnamed "free concentrations" [A] and [H] based on two unnamed "free concentrations" [A] and [H].

$$[HA] = \beta 1[A][H],\ [H_2A] = \beta 2[A][H]_2 \text{ and } [OH] = Kw[H]{-1}$$

The concentrations of "complexes" are therefore estimated using equilibrium constants and unbound concentrations. Standard equations exist for all systems containing reagents A and B. Three or more reagents can be added to this with ease.

3.9. THE USES OF CHEMICAL EQUILIBRIUM

Chemical processes are stable because of chemical equilibrium. Without chemical equilibrium, fluctuating reaction rates would render production impossible (Morgan, 2012).

- **Predicting reaction direction**

Using chemical equilibrium, it is possible to predict the outcome of a process. We can estimate whether the reaction will favor reactants or products by comparing the initial concentrations of reactants and products to the equilibrium constant (K). This knowledge is necessary for the planning and optimization of chemical processes. Regularly assessing the direction of reversible reactions is required. The synthesis of commercial ammonia from nitrogen and hydrogen must be optimized. Predicting the response condition is vital. In chemistry, equilibrium requires equilibrium. It occurs when the concentrations of the reactant and product remain unchanged. As the flavors reach an ideal equilibrium, the reaction time begins to decrease. Do not be fooled; the response is still occurring. Now, both directions are traveling at the same speed.

Understanding chemical equilibrium is enhanced when we realize that "K" represents the equilibrium constant. Calculating the reactant-product equilibrium. By comparing the initial ingredient concentrations to the equilibrium constant, we can estimate which side of the balance will weigh more. If K is extremely high, the response favors product production. If K is less than one, the balance tips in favor of the reactants. A reaction with K close to 1 indicates a balanced reaction in which the amounts on both sides are substantial (Storey, 2011).

What is this information's significance? Consider yourself a skilled caterer preparing a new dish. You desire a dinner with the optimal flavor balance. In some instances, chemical equilibrium permits the prediction of the outcome of a process. It's as if you have a sixth sense that informs you whether to adjust the heating temperature or the ingredients to achieve equilibrium. To achieve flavor harmony, you can plan and create dishes with balance. In addition to cookery, chemical equilibrium has numerous additional applications. Using equilibrium, engineers can improve chemical production operations. By adjusting the temperature, pressure and concentration, they can alter the equilibrium. It is comparable to establishing the optimal environment to increase product output while decreasing waste (Morgan, 2012).

Environmental chemistry requires chemical equilibrium. Globally, pollution is inflicting devastation. Using equilibrium, scientists can anticipate atmospheric responses. This knowledge contributes to the development of methods to prevent hazardous impacts and restore ecological balance in our fragile environment. The chemical equivalent of balancing a dish or the natural world is chemical equilibrium. It is crucial to comprehend how reflexes decline while maintaining their speed. Using the equilibrium constant, we can predict the direction of a reaction and modify the surrounding environment. This information may be used to establish, improve and balance chemical processes in the kitchen and elsewhere.

- **Optimizing Reaction Conditions**

Chemical equilibrium enhances the optimal conditions for a reaction. Temperature, pressure and reactant and product concentrations can all influence equilibrium, allowing for the optimization of product yield or the regulation of reaction size. Consider a chemical reaction with a useful byproduct. Understanding chemical equilibrium can assist in optimizing product production. It is comparable to providing a plant with the proper amount of water and sunlight. By adjusting reactant concentrations, it may be possible to increase product output (Storey, 2011). Controlling the temperature optimizes reaction conditions. Consider simmering sugar-infused water in a kettle. When heated, sugar dissolves more rapidly in water. Nonetheless, mix the sugar as thoroughly as feasible. Increase the temperature to achieve your objective and enhance solubility. It is similar to regulating the garden's climate for optimal growth. In gas processes, pressure influences chemical equilibrium. Consider the movement of gas molecules within a container. Molehills are compressed due to pressure. The equilibrium may favor the side with fewer gas molecules. It is analogous to compacting a garden to enhance plant interaction.

Concentration enhances the conditions of a reaction. Changes in the reactant-to-product ratio can affect the reaction. Increasing the concentration of reactants adds "ingredients" to the process. Reduce its concentration or increase the number of competing reactants to inhibit product formation. Comparable to monitoring and adjusting plant fertilizer to influence growth. Understanding chemical equilibrium enables you to modify these elements to enhance reaction outcomes. You can cultivate a response in the manner of a gardener. As a horticulturist, you can increase production while reducing reactivity (Morgan, 2012). In numerous industries, chemical equilibrium is indispensable for optimizing reaction conditions. It assists engineers and chemists in streamlining industrial processes to increase product yields and reduce waste. An excellent analogy for increasing output while conserving resources is gardening. This comprehension can benefit the environment. Understanding equilibrium enables scientists to mitigate potential detrimental consequences. They can contribute to environmental remediation efforts or limit the release of pollutants. It is comparable to being an environmentally conscious horticulturist who promotes biodiversity and reduces environmental damage. Chemical equilibrium improves reaction conditions. By adjusting temperature, pressure and concentration, reactions can be made as effective as possible. As a master cultivator, you can construct response settings, maximize product yields and manage reaction extent. By comprehending chemical equilibrium, scientists and engineers can create more profitable and environmentally favorable systems (Storey, 2011).

- **Solubility and Precipitation**

The state of equilibrium influences solubility and precipitation. The solubility product constant (Ksp), which represents the equilibrium between a solid and its ions in a solution, can be utilized to predict whether a chemical will dissolve or precipitate. Dynamic equilibrium necessitates solubility. In chemical equilibrium, a solid substance coexists with its solution. The solid may disintegrate, dissociate, or react with an acid or alkali in the solvent. Each equilibrium constant responds differentially to changes in temperature. Solubility equilibrium is required for pharmaceutical, environmental and other applications.

- **Equilibrium of Solubility**

As with any other process, the dissolution of salt in water is governed by chemical equilibrium principles. In several ways, these equilibria differ from the more generic ones provided in the chemical equilibrium lesson set. Examples include the common ion effect, pH and solubility and supersaturation (Morgan, 2012).

- **Ionic Equilibria in Solid Solutions**

As the first step in our investigation of solubility and complexation equilibria, those related to the formation of complex ions, we develop quantitative modeling approaches for the dissolution and precipitation of ionic compounds in aqueous solution. An equilibrium constant expression can be used to express ion concentrations in equilibrium with an ionic material, such as acid-base equilibria.

- **Precipitation and the Solubility Product**

Precipitation and Dissolvability Product In aqueous solutions, cations and anions produce precipitates or intractable ionic substances. Common ionic solid solubility criteria influence the occurrence of this reaction. Before calculating the product state and constructing a net ionic equation, solubility criteria must be verified because not all water reactions result in precipitates (Storey, 2011).

- **The Effects of pH on Solubility**

pH influences the solubility of numerous substances. An anion is a weak acid conjugate base that can be protonated and is present in a large number of infrequently soluble compounds. This section examines the relationship between pH and the solubility of simple binary compounds such as oxides and sulfides, which are both strong bases.

- **Complex Ions and Solubility**

In the presence of complex ions with a high Kf, salts with restricted solubility may become more soluble. A complex ion is formed when a primary metal ion combines forces with one or more ligands, molecules, or other ions containing at least one unpaired electron. Small, highly-charged metal ions are the most common source of Lewis acids and complex ions. Kf is the equilibrium constant for the production of complex ions (Morgan, 2012).

- **Precipitation of Selected Ions**

Qualitative analysis can be used to identify metal ions in complex mixtures of metal ions. By precipitating a few metal ions at a time, it is possible to isolate and identify over 20 common metal cations from a solution. The precipitation stages lose their selectivity.

- **Acid-base chemistry**

Understanding equilibrium ideas is essential for understanding acid-base chemistry. We can measure the degree to which acids and bases ionize in solution using the acid dissociation constants (Ka) and base dissociation constants (Kb). This information is useful for estimating the pH of solutions and comparing the relative strengths of acids and bases. Acid and base partners must balance each other out in acid-base reactions. Bases are the spectators who can take in a positively charged particle known as a proton, whereas acids are the actors who can send it away. What fascinates us about acid-base reactions is this proton exchange (Morgan, 2012).

These protons begin to be released when an acid starts to dissolve in water. Acid dissociation is the name of this procedure. The acid dissociation constant (Ka), a measure of how much an acid releases proton, may be used to determine how much an acid releases proton. It functions as a gauge for how freely the acid distributes its protons. Let's use vinegar, which contains acetic acid, as an example. Acetic acid forms hydronium ions and acetate ions when it dissolves in water by giving a proton to the water molecules. How many acetic acid molecules are changed into these ions is indicated by the acetic acid's Ka value. It's comparable to knowing how many dancers join the stage and begin to move (Storey, 2011).

The base dissociation constant, Kb, can be used to determine how willingly bases will take protons. It reveals how easily a base will take on protons and ionize. Bases are prepared to take protons and join the acid-base reaction, much like performers eager to join a dance. We can determine the relative strengths of acids and bases by knowing the Ka and Kb values. An acid is said to be strong if it has a high Ka value because it rapidly transfers protons. On the other hand, a strong base is one with a high Kb value because it receives protons easily. It's similar to identifying the leading dancers in a group who have the most alluring presence and motions. Let's now discuss pH. A solution's acidity or basicity is determined by its pH. It functions as a scale that indicates how basic or acidic something is. From 0 to 14, the pH scale has a neutral value of 7 (like pure water). Values under 7 denote acidity, whereas those over 7 denote basicity. A solution's pH is correlated with how many hydronium ions (H3O+) are present in that solution. The concentration of hydronium ions rises as an acid dissociates, increasing the acidity and lowering the pH of the solution. On the other hand, when a base takes protons, it reduces the amount of hydronium ions in the solution, increasing its basicity and pH. It's similar to perceiving the intensity of dancing motions, with acidic performances resulting from high proton concentrations and basic performances resulting from low proton concentrations (Morgan, 2012).

We can forecast how compounds will behave in solutions by comprehending acid-base equilibrium. We can gauge an acid's or base's power and foretell how they will interact by knowing the Ka and Kb values. This information is useful in many real-world situations. For instance, understanding the acidity of components is essential for producing the correct tastes in the food and beverage business. To improve the flavor of a meal, a chef may employ various acids, such as citric acid from lemons or lactic acid from yogurt. They may harmonize the tastes and provide a delicious meal by balancing the acid dissociation constants (Storey, 2011).

Understanding acid-base equilibrium is essential for maintaining the body's normal physiological balance, according to medicine. For example, for our cells to operate correctly, our blood must maintain a precise pH range. This information is crucial to the diagnosis and treatment of diseases like acidosis and alkalosis when the pH balance is off. Equilibrium theories and acid-base chemistry help us to understand how acids and bases behave in solutions. We may measure the degree of ionization and ascertain the relative potencies of acids and bases using the acid dissociation constant (Ka) and base dissociation constant (Kb). This information is crucial for determining the pH of solutions and is used in a variety of industries, including food, medical and environmental science. It's similar to knowing the tastes and movements of the dance between acids and bases, which deepens our grasp of the chemical world.

- **Reaction kinetics**

Chemical equilibrium and reaction kinetics are closely related concepts. Reaction kinetics focuses on the rates at which reactants are transformed into products, whereas equilibrium addresses the overall balance between reactants and products. Understanding the variables that affect reaction rates and the methods by which equilibrium is achieved is aided by the study of chemical equilibrium. Reaction kinetics and chemical equilibrium go hand in hand. process kinetics looks at how soon the balance between reactants and products is reached, whereas equilibrium discusses the equilibrium between reactants and products in a process. The time it takes for a reaction to achieve equilibrium might vary based on several circumstances, just like when preparing a meal (Morgan, 2012). The reaction rate is a crucial idea in reaction kinetics. Consider sautéing some veggies in a skillet. The amount of heat used, the size of the veggies and the type of cooking surface all affect how quickly the vegetables cook. Similar to this, the pace of chemical reactions relies on elements including temperature, reactant concentration, surface area and catalyst presence (Storey, 2011).

The kinetics of reactions are greatly influenced by temperature. The rate of a chemical reaction can be sped up by raising the temperature, much as increased heat can speed up cooking. This is because greater temperatures give reactant molecules more energy, increasing the possibility of successful collisions and reactions. Another important element in reaction kinetics is concentration. Consider that you are producing cake batter and discover that increasing the sugar by twofold makes the batter sweeter. The frequency of molecular collisions rises in chemical reactions as reactant concentration rises, speeding up the reaction rate. It's like having more of the required ingredients on hand, making the cooking process go more quickly.

In reaction kinetics, the surface area is also significant. Consider how the flavor of garlic spreads more rapidly when it is finely diced as opposed to when it is intact. Similar to this, grinding or finely splitting reactants increases their surface area, allowing for greater molecular interaction and accelerating chemical reactions. Catalysts can also have a substantial impact on the kinetics of a process. Catalysts accelerate the rate of reactions without being consumed themselves, just as adding yeast to dough speeds up the bread-making process. They offer a different route for the reaction, decreasing the energy barrier needed for it to happen. As a result, the reaction proceeds more quickly and more reactant molecules can engage in successful collisions (Morgan, 2012).

In many domains, it is essential to comprehend how chemical equilibrium and reaction kinetics relate to one another. By comprehending the speeds at which reactions occur and determining the most effective settings, reaction kinetics, for instance, aids scientists in the pharmaceutical industry in optimizing the manufacture of medications. Reaction kinetics is a tool used in environmental research to investigate how contaminants degrade. Scientists can devise plans for the remediation of dangerous compounds by understanding the reaction rates and calculating how long it takes for harmful substances to decompose in the environment. Reaction kinetics aids in our comprehension of the complexity of biological systems. Biochemical reactions rely heavily on enzymes, which are biological catalysts. We may unravel the complex systems governing the chemical reactions in our bodies by comprehending their kinetics. Studying reaction kinetics is similar to watching food being prepared and learning how ingredients change into finished goods over time. It gives information on the catalysts, surface area, temperature, concentration and other variables that affect reaction rates. This information aids in response optimization, the creation of effective procedures and a greater comprehension of the mechanics behind equilibrium formation. Understanding reaction kinetics enables scientists to regulate reactions and unlock the mysteries of the chemical universe, much as a great chef may manage the cooking process to produce the ideal dish (Storey, 2011).

- **Industrial processes and environmental chemistry**

Chemical equilibrium is crucial to the production of minerals, drugs and other industrial products. The optimization of reaction conditions is made possible by an understanding of equilibrium circumstances, which increases productivity and decreases waste. Equilibrium principles are useful in environmental chemistry for understanding reactions like acid rain, atmospheric chemistry and chemical equilibria in natural waterways. Chemical equilibrium serves as a guiding concept in the production industry. Scientists and engineers employ equilibrium conditions to fine-tune their reactions, much as a competent conductor makes sure that every instrument performs in unison. To get the optimum results, they carefully modify variables like temperature, pressure and reactant and product concentrations. Haber's technique uses it to make ammonia from it. Nitrogen and hydrogen gas makes ammonia here. Low temperature, high pressure and iron catalyst increase ammonia production. It makes contact-processed sulfuric acid: This technique oxidizes sulfur dioxide to sulfur trioxide (Morgan, 2012).

One of the important variables that may be managed to enhance industrial operations is temperature. Consider yourself employed at a steel plant, where iron is converted into robust steel. Engineers can determine the ideal temperature range that permits the intended chemical processes to happen effectively by understanding equilibrium. Similar to establishing the ideal balance between the risk of overheating or equipment damage and the heat required for the reactions to continue. Another variable that may be changed to enhance industrial operations is pressure. Imagine yourself at a petrochemical facility, where crude oil is converted into useful materials like plastic and fuel. Engineers can control the equilibrium of different reactions and encourage the creation of desired products by modifying pressure. To increase yield and process efficiency, it's like finding the ideal pressure to force reactions in the desired direction. In industrial processes, reactant and product concentrations are also crucial. Imagine yourself working on a novel medicine in a pharmaceutical facility. Scientists can optimize the reaction conditions to maximize the yield of the intended product and reduce the development of undesirable byproducts by carefully managing the amounts of reactants. It's comparable to making certain that a recipe contains the ideal proportion of components to yield the tastiest food (Storey, 2011).

Scientists and engineers may improve the efficiency and environmental friendliness of industrial operations by studying chemical equilibrium. They can eliminate waste, consume less energy and enhance overall process economics by optimizing reaction conditions. It's similar to discovering ways to simplify the production process so that fewer resources are used and less environmental damage is caused. Chemical equilibrium, however, is important in environmental chemistry as well as industrial processes. Consider interactions like acid rain, in which contaminants from industrial processes mix with elements of the atmosphere to form acidic precipitation. The magnitude and impact of these reactions may be studied and predicted by scientists with the use of equilibrium ideas. In atmospheric chemistry, where reactions involving gases and particles cause air pollution and climate change, equilibrium ideas are also relevant. Scientists can evaluate the impact of pollutants and create plans to lessen their impacts by knowing the equilibrium between various chemical species in the atmosphere. It's similar to comprehending how the many elements in the air combine to produce the meteorological conditions we experience. Equilibrium theories aid in the understanding of chemical equilibrium in natural waters in the field of environmental chemistry. In aquatic habitats, which range from rivers and lakes to oceans, different chemical reactions occur. Understanding these equilibria aids in determining the quality of the water, researching nutrient cycles and even creating techniques for cleaning up polluted streams. It is comparable to understanding the intricate dance of chemicals in water bodies and how it affects the ecology. Both industrial operations and environmental chemistry depend on chemical equilibrium. It enables engineers and scientists to improve reaction conditions, boosting productivity and decreasing waste in industrial operations. As well as analyzing environmental processes like acid rain, atmospheric chemistry and chemical equilibria in natural waterways, equilibrium principles are crucial. We may build a more sustainable future where industrial operations are effective, ecologically responsible and our natural ecosystems are preserved by comprehending and utilizing the principles of balance (Storey, 2011).

CHAPTER 4

Chemical Kinetics with Enzymology

Contents

4.1. INTRODUCTION

Chemical kinetics examines reaction rates and the factors that affect them. Chemical reactions in living things are speed up by enzymes, which are biological catalysts. Proteins are made of amino acid chains. A core carbon atom, an amine group, a carboxyl group, a hydrogen atom and a unique side chain are all present in each amino acid. Protein structure is determined by amino acid interactions and sequence. Several of these interactions are influenced by hydrogen ion concentration. As a result, amino acid bonds are strengthened, giving the enzyme its three-dimensional shape. The three-dimensional structure of enzymes allows them to attach to substrates and carry out chemical reactions. Altering the interactions between amino acids and the structure of an enzyme can change its activity. Since enzymes accelerate chemical processes without eating them, they are essential to chemical kinetics. The process is accelerated by enzymes by lowering the activation energy (Knowles, 2014).

The Gibbs free energy function (G), which applies chemical thermodynamics to biological catalysis, operates best under physiological conditions of constant temperature and pressure. The free energy change of a chemical reaction can have an impact on its spontaneity under specific circumstances. Equilibrium is governed by G°. G°′, the biochemical standard free energy change, is used in biochemistry. Keep in mind the following information about free energy:

- At constant temperature and pressure, a thermodynamically spontaneous process exhibits a negative Gibbs free energy change ($\Delta G < 0$). The G value for a spontaneous process must be negative (Srinivasan, 2021).
- Equilibrium is $\Delta G = 0$. As a function of the concentrations of the product and reactant, this condition minimizes the free energy G. The equilibrium is dynamic but does not modify the composition since G has fallen to its lowest level.
- If $\Delta G > 0$, a process cannot happen by itself. A process with a negative net G must be coupled to the active process.
- The equilibrium constant is K′ and the biological standard free energy change for the process is $\Delta G^{\circ\prime} = -RT$.
- If $\Delta G < 0$ the rate of a process is unknown. Response rates are explained by transition state theory. Biochemical equilibrium is accelerated by enzymes.

Erroneous reactant and product species concentrations may be used to constrain in vivo activities like metabolic processes when evaluating G. When there is a paucity of information on metabolite concentrations in vivo, comparative G° values are used.

A kinetic rate constant (small k) can be related to the fictitious free energy required to reach the transition state (G), much as in equilibrium thermodynamics (big K). Kinetics makes use of the transition state hypothesis.

The following fundamental idea of enzymatic catalysis takes into account kinetics and thermodynamics. Enzymes specifically inhibit process G after producing an ES complex to hasten equilibrium. A crucial mechanical idea is preferential binding in a transition state (ES) complex. In order to comprehend how the enzyme active site binds cognate substrates and selectively reduces the energy of the chemical transition state(s), we study enzymes and their complexes with substrates or inhibitors (Knowles, 2014).

4.2. TRANSITION STATE THEORY

Thermodynamics enables us to quantitatively relate the free energy of activation, G, to the speed of a chemical process.

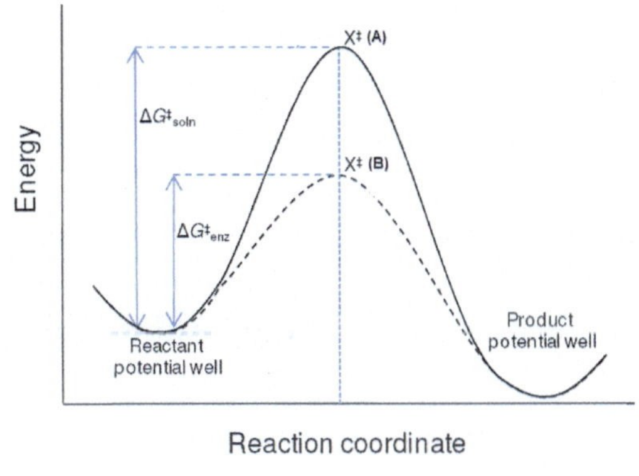

Figure 4.1. Transition state theory model of catalysis.

Source: Quaye, Osbourne. (2009). On the Preorganization of the Active Site of Choline Oxidase for Hydride Transfer and Tunneling Mechanism. Scientific Figure on ResearchGate. Available from: https://www.researchgate.net/figure/Transition-state-theory-model-of-catalysis-X-A-is-the-transition-state-for-a_fig4_268304262 (accessed on 22 September 2023)

The rate enhancement an enzyme can achieve by reducing G by a certain amount may then be calculated. The reaction's velocity (rate) is proportional to the reactant concentration and the exponential of the ratio (G/RT), assuming that the reactant and reaction transition stage are in equilibrium. An enzyme may speed up by 10 times at 25 °C for every 5.71 kJ/mol (1.36 kcal/mol) drop in G (Figure 4.1).

The transition state corresponds to the species with the greatest energy throughout the chemical pathway. In addition to partial bond cleavage and formation, it possesses reactant and product properties. This is another name for the transitional phase, which is also known as the activated complex.

In order for a chemical reaction to occur, the activation energy (Ea) must be overcome. This is because of the difference in energy between the reactants and the transition state. As the activation energy increases, the rate of the reaction decreases (Srinivasan, 2021).

The rate constant (k), also known as the proportionality constant, relates the rate of a chemical reaction to the concentration of reactants. The Arrhenius equation illustrates how the rate constant varies with temperature: In the equation $k = Ae(-Ea/RT)$, A is the preexponential component, R is the ideal gas constant and T is the Kelvin temperature.

The rate of a chemical reaction is determined by the rate at which reactant molecules overcome the activation energy barrier and reach the transition state. Measuring the concentration of a reactant or product per unit of time is a common technique for characterizing the rate of a chemical reaction.

According to transition state theory, the equilibrium position of a reaction is dictated by the location of the transition state along the reaction coordinate. The forward reaction is thermodynamically more favorable when the energy of the transition state is greater than the energy of the reactants.

Catalysts speed up chemical reactions by supplying a new reaction pathway with a lower activation energy. In order to accomplish this, catalysts either initiate a new reaction pathway or stabilize the transition state. As biological catalysts, enzymes frequently operate in accordance with transition state theory principles. The reaction coordinate is a fictitious route depicting the progression of reactants to products in a chemical reaction. It describes each bond breaking and bond forming transition that takes place throughout the reaction. The transition state is located along the reaction coordinate at its highest point. The Hammond postulate, also referred to as the Hammond-Leffler postulate, states that the species (reactants or products) produced by a chemical reaction are more energetically similar in their transition state. In other words, the structure of the transition state varies based on whether it is energetically closer to the reactants or the products (Knowles, 2014).

Free energy diagrams, also known as potential energy diagrams and reaction coordinate diagrams, depict the energy transformations that occur during chemical reactions. Along the reaction coordinate, they display the relative energies of the reactants, products and transition state. The difference between the reactant and transition state energies is the activation energy.

Analogues are substances whose structure and properties resemble those of the transition state. Because they are designed to bind to enzymes or catalysts with high specificity and affinity, these compounds can be used to investigate enzyme function as inhibitors or probes. Analogs of the transition state are essential for comprehending the kinetics of enzymes and developing new medications (Srinivasan, 2021).

In multiphase, complex processes, the rate-determining phase affects the overall reaction rate. Typically, the transition state with the greatest energy barrier occurs during this procedure. For designing efficient catalysts and optimizing reaction conditions, it is essential to comprehend the rate-determining phase. The Marcus theory is used to incorporate electron transport pathways into the transition state theory. The energy difference between reactants and transition states, as well as the reorganization energy associated with the rearrangement of solvent molecules and the surrounding environment, must be utilized to describe the rate of electron transfer processes.Theorems from quantum mechanics can be utilized to develop transition state theory, which provides a deeper understanding of the behavior of atoms and molecules at the subatomic level. Quantum mechanics calculations can be used to evaluate the energy landscapes and calculate the reaction rates of complex systems. In addition to predicting and explaining the impact of temperature, concentration and catalysts on reaction kinetics, the transition state theory provides a chemical explanation for reaction rates. It provides a framework for organizing and enhancing chemical methods in a variety of fields, such as chemical synthesis, drug discovery and materials research (Srinivasan, 2021).

4.3. CATALYSIS AND ENZYME MECHANISMS

Multiple primary steps and transition phases are involved in reactions that enzymes catalyze. The study of enzyme mechanisms is crucial to the field. The chemical and structural diversity of proteins makes them effective enzymes. As we'll see, non-protein cofactors broaden the chemical toolbox of enzymes. Even though the players in mechanistic models of enzyme operation are quite diverse, many enzyme studies have several characteristics.

Reversely, we may apply a few mechanistic principles to the diverse spectrum of enzymes and processes they catalyze as a unifying biochemistry approach. Enzymes aren't different; they're just better, as one well-known enzymologist put it. It helps to understand how organic chemistry works. On carbonyl compounds and their derivatives from carboxylic acids, several metabolic activities rely. The most common mechanistic types of catalysis are general acid/base, covalent and metal ion catalysis. Due to the chemical diversity of the twenty amino acids, proteins can catalyze reactions through a number of first-two processes. In many organic chemistry reactions, covalent catalysis includes nucleophilic attack. In catalytic reactions, universal acids like -COOH and ImH+ (protonated imidazole) serve as proton donors whereas general bases like -COO, Im and -NH2 serve as proton acceptors. Ser and Cys' -OH and -SH groups are examples of nucleophiles found in protein side chains (Figure 4.2) (Knowles, 2014).

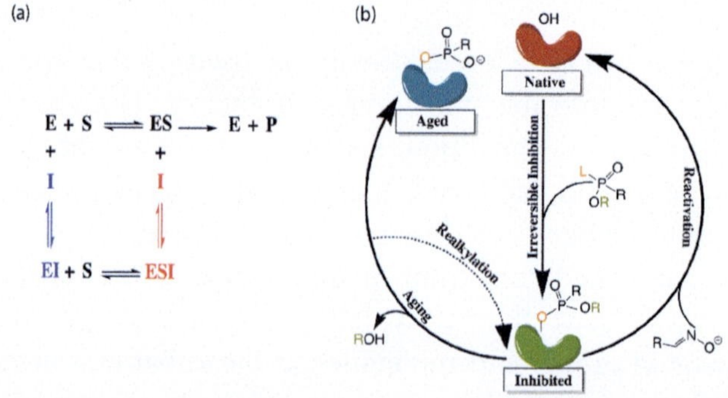

Figure 4.2. (a) Depiction of the enzyme catalytic mechanism (black) and reversible inhibition mechanisms including competitive inhibition (blue), uncompetitive inhibition (red) and noncompetitive inhibition (both blue and red). (b) Representation of irreversible (covalent) inhibition by organophosphorus ligands and the reactivation mechanisms to revert the aged enzyme (blue) back to the inhibited complex (green).

Source: Boer, Danna & Nguyen, Nguyet & Mao, Jia & Moore, Jessica & Sorin, Eric. (2021). A Comprehensive Review of Cholinesterase Modeling and Simulation. Biomolecules. Scientific Figure on ResearchGate. Available from: https://www.researchgate.net/figure/a-Depiction-of-the-enzyme-catalytic-mechanism-black-and-reversible-inhibition_fig2_350905074 (accessed on 22 September 2023)

4.4. COFACTORS

Protein groups are not the only need for electrophilic catalysis. Cofactors are used by enzymes to increase their catalytic efficiency. Cofactors serve as "accessories" for catalysis in many enzymes. Further distinctions are made between "small" organic cofactors and metal. A coenzyme that is tightly bonded (typically covalently) and that the enzyme is unable to exchange is known as a prosthetic group. Cofactors for enzymes serve a number of purposes. They may act as catalysts, assisting enzymes in carrying out chemical reactions that proteins struggle with. In other situations, coenzymes move acyl groups, one-carbon units, or electrons. Although their modular nature—exchanged as part of the catalytic turnover, recycled and used by a number of enzymes—and their connection to vitamins (from which they are created) differentiate them from typical substrates in this situation, co-substrate may be more applicable. Cofactors with metal ions catalyze. Metal ion catalysis is demonstrated by Zn2+ in carbonic anhydrase. Co-substrates and prosthetics are capable of transporting chemical groups (NADH, FADH) or electrons. Acyl groups are transported via coenzyme A. Amino groups are accepted by PLP. These cofactors are examples of enzymes (Knowles, 2014).

4.5. ENZYME ACTIVE SITES AND ENZYME-SUBSTRATE (ES) COMPLEXES

The active site is a three-dimensional cleft or "pocket" formed by the residues of the enzyme's tertiary structure. Due to the mismatch between the primary and tertiary structures, an enzyme-active site can be formed by residues far apart in the sequence. By interacting directly with substrates, enzymes catalyze reactions. Before the direct structural determination of enzymes and their complexes, several lines of evidence supported the formation of enzyme-substrate complexes. Saturation kinetics are a feature of all enzymes. Reaction velocity peaks as substrate concentration rises, indicating that the catalytic sites of the enzyme are full. After substrate addition, enzyme modifications may be inferred using UV-visible or fluorescence spectroscopy. Several structural illustrations of how enzymes detect and bind their substrates have been produced by X-ray crystallography of enzyme-substrate or enzyme-substrate analog complexes (Srinivasan, 2021). Since most enzymatic reactions include more than one substrate (reactant), take into account the consequences of proximity and orientation. As we will see, if an enzyme catalyzes a process involving substrates A and B, a ternary enzyme-substrate complex (EAB) may form. By merely combining the substrates at the active site, an enzyme can enhance the products of the reaction A + B. This significantly increases the reactants' effective concentration, which, according to chemical kinetics, accelerates reaction rates. Additionally, it is important to arrange the proximity of A and B so that the answer may be oriented in the best way possible. The significance of orientation is demonstrated by the geometry of the organic SN2 reaction. The assault by the nucleophile is targeted. In order to catalyze a reaction, enzymes might bind substrates in the active site in a specific way.

4.6. BINDING FORCE

Consider the binding energy of a substrate, the energy change during the formation of the ES complex and the question of whether an enzyme can use a portion of the binding energy to be better suited to binding a transition state species in order to comprehend the preferential binding of the transition state. An excellent example of binding energy is lysozyme. Numerous energetically advantageous interactions exist on its broad substrate, allowing for strong binding. Instead of binding to its substrate, lysozyme hydrolyzes it. The substrate is altered by enzyme interactions to approximate the reaction transition state.

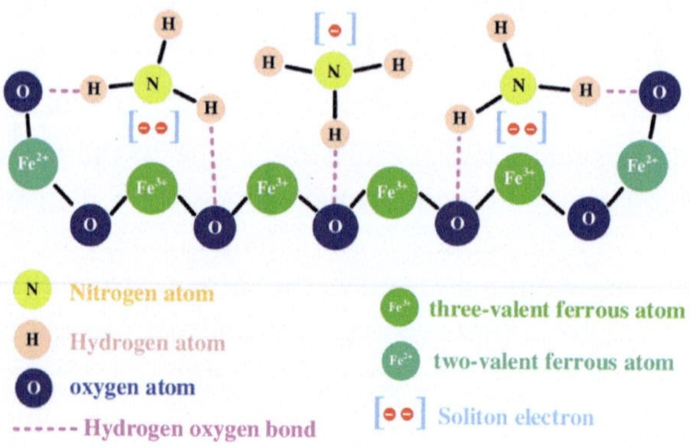

Figure 4.3. Model of the intermolecular binding force.

Source: Guo, Tao & Zhou, Tianhao & Tan, Qiulin & Guo, Qianqian & Lu, Fengxiang & Xiong, Jijun. (2018). A Room-Temperature CNT/Fe3O4 Based Passive Wireless Gas Sensor. Sensors. Scientific Figure on ResearchGate. Available from: https://www.researchgate.net/figure/Model-of-the-intermolecular-binding-force_fig1_328395141 (accessed on 22 September 2023)

This is another approach to look at the idea of reducing the energy of reaction transition states. Lysozyme binds to the transition state more strongly than its substrate, which makes a catalytic difference. The enzyme-substrate combination gives up some binding energy in order to bind a substrate with higher energy (Figure 4.3) (Knowles, 2014).

Following an introduction to enzymes, we examine case studies to comprehend how they achieve catalytic power and substrate selectivity by employing strategies that fit our paradigm. Catalytic power and specificity are influenced by form and electrostatic complementarity. Specificity and reaction activation energy are dependent on the complementarity of the substrate and transition state. In contrast to the vast ensemble of potential random coil configurations, enzymes, like proteins, take a folded, functional conformation that is energetically advantageous. However, this does not imply that there is just one such conformation. In fact, the ability of an enzyme to adopt a variety of conformations is frequently what determines how active the enzyme is and there is a wealth of data to support this. Natural selection has produced incredibly effective enzymes thanks to the ability of proteins to alter shape and generate exact electrostatic conditions (Srinivasan, 2021).

4.7. THE CONCEPTS AND FACTORS AFFECTING CHEMICAL KINETICS AND ENZYMOLOGY

4.7.1. Enzyme-Substrate Complex

An enzyme-substrate complex is produced when enzymes bind to certain substances referred to as substrates. It is feasible to transform raw materials into finished products at this degree of complexity. The binding of the substrate to the enzyme's active site is frequently one of the most significant rate-determining steps in the whole process. When an enzyme makes complete contact with its substrate, a molecule known as the enzyme-substrate complex is produced. When an enzyme is dormant, its structural makeup is somewhat altered. When the substrate approaches the active site, it goes through a conformational shift, or change in structure. The active site of an enzyme is the region where a chemical bond with the substrate can form. This shapeshift can push two or more substrate molecules away or together. Enzymes must be able to respond quickly enough to be useful for the majority of the functions that cells require to live. The organism's DNA provides specific instructions for these enzymes (Knowles, 2014).

The enzyme-substrate complex is necessary for a number of reasons. The complex of the enzyme substrate is first fleeting. This demonstrates that after being altered, the enzyme can no longer attach to the substrate. When the waste products are removed, the enzyme is ready to accept a new substrate molecule. Each cell only requires a small quantity of the enzyme because it may function millions of times. Enzymes are complex molecules that operate like miniaturized machines. The contact, bending and twisting of these various amino acids with one another results in a complicated structure formed of a long chain of amino acids. These components can be employed as hinges, wedges and in a number of other ways to speed up certain reactions. Diverse mutations produce enzymes of comparatively diverse kinds. Beneficial mutations modify the production of the enzyme's products or overall activity via altering the enzyme substrate complex. For this modification to be advantageous, the organism has to be motivated to procreate more (Srinivasan, 2021). In addition to the -ase suffix, which identifies them as enzymes, the names of enzymes typically come from the substrate on which they function. Which molecules an enzyme can bind to depends on how specifically it interacts with the substrate. Because they are unable to catalyze the reaction that the enzyme is trying to carry out, some molecules with the same structure as the substrate may become stuck in the active site. The competitive inhibitor binds to the bent substrate-enzyme complex and prevents the enzyme from continuing to operate. Some inhibitors imitate the substrate, while others alter the enzyme in various ways to prevent the formation of the enzyme-substrate complex (Knowles, 2014).

Amylase and amylose are two examples of substrate enzyme complexes. Amylose is a complex sugar produced by plants. The salivary enzyme amylase breaks down amylose. Amylase requires one molecule of the cofactor water and one molecule of the substrate amylose in order to put together an enzyme substrate complex. The complex lengthens the time it takes for a reaction to begin by greatly reducing the energy required to initiate it. Without enzymes like amylase, it would take millions of years to break down a normal sugar molecule.

Because enzymes are so important to the metabolism of the food we consume, our bodies produce enzymes for practically every type of food that they have evolved to digest. Because the enzymes have not had enough time to mature, newly consumed meals are not completely digested. For instance, the current diet of processed foods, which is high in nutrients only available through the routes used to store fat, is to blame for the obesity epidemic. As a result, a large portion of the population struggles with weight-related health issues. Many dietitians advocate for natural, whole-food plant-based diets because they support the enzymes that our bodies have evolved to naturally produce (Srinivasan, 2021).

Allosteric controls the activity of enzymes

In order for the enzyme substrate complex to form fast, it is occasionally necessary to "turn off" an enzyme in order to conserve energy or other resources. This regulates a huge number of enzymes to produce the perfect quantity of energy and goods. The molecule adenosine triphosphate (ATP), which supplies energy to cellular processes, is one of the most important locations for this. ATP inhibits a large number of the enzymes that are a part of the route that produces ATP. When too much ATP is produced, the enzyme stops functioning. This is referred to as self-regulation or feedback suppression. Adenosine diphosphate (ADP), an ATP whose phosphate group has been used to supply energy to a method or reaction, can also be used to rebuild enzymes. In this manner, several physiological processes are regulated and the precise molecules are needed to produce the enzyme substrate complex. Cofactors are components in the diet such as vitamins, minerals and other inorganic elements that help activate enzymes (Knowles, 2014).

4.7.2. Effect of Temperature

Like all other chemical reactions, enzymatic processes are affected by temperature. As a result of enhanced molecular mobility, the reaction rate typically increases with temperature. Enzymes may become inactive due to denature, which can occur at excessive temperatures. While enzymes can function more swiftly at higher temperatures, if the temperature rises too much, the enzyme will cease functioning. Enzymes require incredibly precise three-dimensional structures in order to function properly. An enzyme loses its shape and ceases to function if the temperature around it rises too much, a condition known as denaturation. At really high temperatures, the majority of enzymes denature. When the temperature reaches the optimal level, the kinetic energy of the enzyme and substrate increases, causing more collisions between the two. The pH and temperature ranges where an enzyme's catalytic activity is at its highest are universal. Two known temperature parameters have an impact on how quickly enzymes catalyze reactions. A couple of these include thermal stability and activation energy.

The results on the relationship between temperature and enzyme activity do not totally support the assertion that activity increases with temperature. A novel model known as the Equilibrium model, which is based on Le Chatlier's concept of fundamental chemistry, helps to the quantitative explanation of enzyme thermal behavior under reaction circumstances by taking into account the enzyme's inactive state and building an equilibrium system that does so (Srinivasan, 2021).

By removing time dependence, which was considered to be crucial to the conventional understanding of enzyme activity vs temperature, the equilibrium model significantly advances the field. When an active enzyme is exposed to different temperature gradients, the equilibrium model states that the existence of an inactive enzyme prevents the active enzyme from becoming completely inactive.

Notably, it is essential to comprehend the differences between the active and inactive forms of the enzyme. In order to reflect an enzyme's inactive state, the equilibrium model completely focuses the enzyme's activity on its active site. The active site of an enzyme can get attached to substances, causing a conformational change that enables the enzyme to restart carrying out its biochemical job.

According to this theory, the only difference between an enzyme's active and inactive states is a conformational change. These enzymes are distinct from denatured enzymes. Enzymes that have been rendered inactive have experienced an irreversible conformational change that prevents them from performing their usual function.

The mechanics of the equilibrium model are straightforward due to the reversibility of the inactive enzyme form. The equilibrium mechanics of the proposed model is consistent with the potential of an enzyme transition from an inactive to an active state. While enzymes can function more swiftly at higher temperatures, if the temperature rises too much, the enzyme will cease functioning.

An enzyme loses its shape and ceases to function if the temperature around it rises too much, a condition known as denaturation. More enzyme-substrate complexes form as a result, enhancing the enzyme's activity. When the temperature exceeds the optimum threshold, the enzyme's active site becomes denatured (Knowles, 2014).

Because the active site has lost its essential structure and is unable to form complexes with substrates, enzyme activity declines. Enzymes gradually lose their activity, even at very low temperatures. Usually, it's recommended to keep enzymes at or below 5 degrees Celsius. At lower temperatures, chemical processes take longer to complete. The majority of enzymes are dormant at low temperatures and dramatically reactivate at higher temperatures, while certain enzymes become inactive when frozen (Srinivasan, 2021).

4.7.3. Internal Energy and Kinetic Energy

The temperature of a system may affect the kinetic energy of the molecules inside it. The frequency of molecule collisions grows with temperature. This is due to the fact that kinetic energy and velocity increase as temperature increases. The space between collisions gets less as the speed increases. More molecules gravitate toward the activation energy as a result, accelerating the processes.

As a result of the faster-moving molecules, there are more collisions between substrates and enzymes. As a result, the system temperature rises as kinetic energy rises and lowers as kinetic energy drops. The intrinsic energy of the molecules will increase as the system's temperature does. Internal energy is the energy of a molecule's internal motions, including translation, vibration and rotation, as well as chemical bonding and nonbonding interactions.

This heat can be partially converted into chemical potential energy. Some of the weak connections that regulate the three-dimensional structure of active proteins may collapse if the chemical potential energy rises to a certain level. The protein may then be rendered inactive by heat denature. High temperatures can therefore slow down an enzyme-catalyzed process because they can denature or inactivate an enzyme or substrate (Knowles, 2014).

4.7.4. Optimum Temperature

Every enzyme has a temperature range in which its reaction rate is the quickest. This maximal temperature is referred to as the optimal temperature for the enzyme. A temperature of roughly 98.6 degrees Fahrenheit (37 degrees Celsius) is preferred by the majority of enzymes. Additionally, a number of enzymes are active at both low and high temperatures. For instance, animals in arid conditions have developed enzymes that work best at higher temperatures, whereas animals in cold settings have evolved enzymes that work best at lower temperatures.

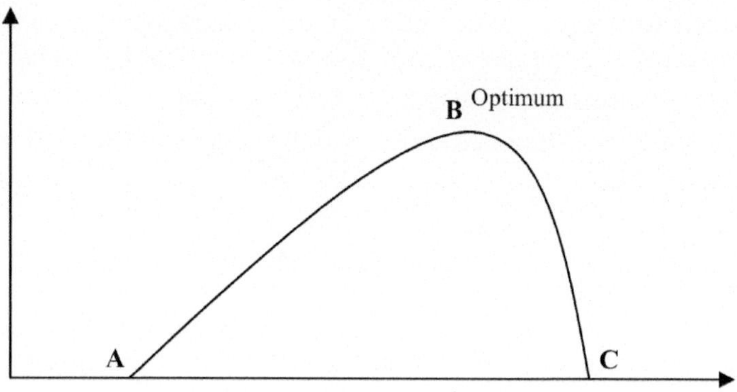

Figure 4.4. The cardinal temperatures for microorganisms (minimum temperature A, optimum temperature B and maximum temperature C).

Source: Niwagaba, Charles. (2023). Human Excreta Treatment Technologies – prerequisites, constraints and performance. Scientific Figure on ResearchGate. Available from: https://www.researchgate.net/figure/The-cardinal-temperatures-for-microorganisms-minimum-temperature-A-optimum-temperature_fig9_253275382 (accessed on 22 September 2023)

Similar to other proteins, enzymes begin to degrade at temperatures over 104 degrees Fahrenheit. As a result, an enzyme's activity range is determined by the temperature at which it begins to function and the temperature at which a protein begins to degrade (Figure 4.4).

4.7.5. Effect of pH

A certain pH range is when enzyme activity is at its highest. Enzyme deactivation or denaturation may occur if the pH is raised over this range. Enzyme sensitivity to pH can provide information about the chemical processes that they catalyze. The activity and/or speed of a chemical reaction are both controlled by the structure of an enzyme. In other words, the rate at which a reaction occurs depends on changes in the structure of the enzyme. The structure of the enzyme is also impacted by the pH of a medium. The charge and shape of the substrate can be impacted by pH as well as enzyme activity. Each enzyme has a specific pH requirement. At this pH, an enzyme in particular catalyzes the reaction more quickly than at any other pH (Knowles, 2014). The pH range of 5 to 9 is generally where enzyme activity peaks. This is so because proteins function in a pH-reflective environment. For several reasons, the pH range of proteins is constrained. The pKa ranges of proteins' carboxyl and amide termini, as well as a number of amino acid residues, are linked to the intracellular pH range. A pH change can thereby protonate or deprotonate a side group, changing the chemical composition of the side group. For instance, improper deprotonation of carboxyl termini may cause them to lose contact with a neighboring subunit, changing the structure of the enzyme. Not to mention, this configuration could lower substrate affinity. Greater pH changes can change how proteins fold, inactivate the enzyme and start irreversible proteolysis. On the other hand, pH changes could enable enzymes to regulate or affect protein activity. Hemoglobin can create a salt bridge when the blood plasma is acidic. T-state hemoglobin is hence less oxidative and more stable. This enhances the delivery of oxygen to muscles that are lacking in it (Srinivasan, 2021).

- The substrate is initially bound by the enzyme.
- The ionization states of the residues of amino acids in the catalytic site of the enzyme.
- The degree of ionization of the substrate.
- Modifications to protein structure (particularly apparent at pH levels below really low levels).

The optimal pH for each protease enzyme varies, with pepsin, another protease enzyme, performing best at an acidic pH and trypsin, another protease enzyme, performing best at a slightly alkaline pH. The pH scale is used to assess a solution's acidity or alkalinity. The pH of a solution is a more accurate representation of the concentration of hydrogen ions (H+) dissolved in a solution and the ion concentration fluctuates with the pH of a solution. By creating new bonds or rupturing ones that already exist, these ions change the structure of the enzymes and, occasionally, the substrate. Towards the end, the enzyme and substrate both go through chemical changes. The enzyme's active site is also changed, making it challenging for the substrate to recognize the enzyme. For further details on enzymes, visit the complex of enzyme substrates.

Take the following scenario into consideration: Lower the pH level below its ideal range to influence the procedure. There will be a change in the enzyme activity or reaction rate from the prior instance. Nothing is possible to occur continuously. When structural elements of the substrate and active site alter, this happens. In order for the chemical reaction to take place, the pH of the solution must be adjusted to a value that is adequate for both the enzyme and the substrate. This enables research into how pH affects enzyme function in real-world situations (Srinivasan, 2021).

4.7.6. Enzyme Inhibitors

Molecules called enzyme inhibitors prevent enzyme activity. Both irreversible and reversible inhibitors fall within this group. The study of enzymatic activity control and the creation of drugs that target particular enzymes benefit from knowledge of the kinetics of enzyme inhibition. Chemicals called enzyme inhibitors attach to enzymes and stop them from working. Because they speed up chemical reactions that change substrate molecules into products, enzymes, which are proteins, are crucial for life. An enzyme speeds up a particular chemical reaction by anchoring the substrate to its active site, a specialized area on the enzyme that speeds up the trickiest part of the process (Knowles, 2014). By sticking to the enzyme's active site and preventing the substrate from binding there or by attaching to another part of the enzyme and preventing it from catalyzing the reaction, an enzyme inhibitor prevents this activity from occurring (or "inhibits" it). Inhibitors of enzymes can bind reversibly or permanently.

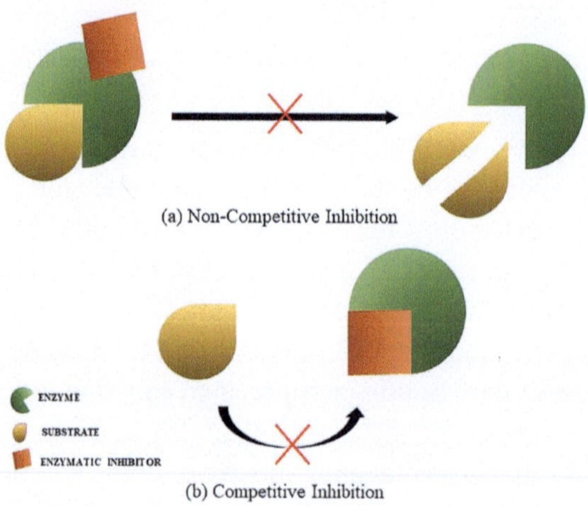

Figure 4.5. Mechanisms of enzymatic inhibition. (a) non-competitive inhibition. (b) competitive inhibition.

Source: Lima, Vanessa & Piuvezam, Grasiela & Maciel, Bruna & Morais, Ana. (2019). Trypsin inhibitors: promising candidate satietogenic proteins as complementary treatment for obesity and metabolic disorders. Journal of Enzyme Inhibition and Medicinal Chemistry. 34. 405–419. Scientific Figure on ResearchGate. Available from: https://www.researchgate.net/figure/Mechanisms-of-enzymatic-inhibition-a-non-competitive-inhibition-b-competitive_fig2_330218908 (accessed on 22 September 2023)

Irreversible inhibitors and the enzyme form a chemical link that inhibits the enzyme from working until the bind is broken. On the other hand, reversible inhibitors bind non-covalently and may spontaneously depart from the enzyme, enabling it to resume its regular action. There are many types of inhibition that can happen as a result of reversible inhibitor binding to the enzyme, the enzyme-substrate complex, or both (Figure 4.5).

All living things need enzyme inhibitors since they typically only inhibit one enzyme at a time and control its activity. For instance, later-formed compounds in a metabolic pathway may inhibit certain enzymes, preventing the formation of extraneous molecules. A cell has to experience this kind of negative feedback in order to stay in homeostasis. Nucleases and proteases are two enzymes that are controlled by enzyme inhibitors. Toxins produced by animals or plants called enzyme inhibitors prevent vital enzymes from working in prey or predators.

A class of medication known as enzyme inhibitors stops the development of pathogens such viruses, bacteria, parasites, or aberrant human enzymes. Two examples include methotrexate, a chemotherapy drug that also treats rheumatoid arthritis and protease inhibitors, which are used to treat HIV/ AIDS. Since anti-pathogen inhibitors commonly only target one enzyme, they are quite selective and, provided no human analogues for that enzyme exist, usually have no negative effects on people. This is typically the case because of the genetic distance between these conditions and people. It takes a little amount of pharmacological enzyme inhibitors to block the enzyme since they often have low dissociation constants. The risk of liver and kidney damage in humans, as well as other unfavorable pharmacological consequences, is decreased when the enzyme inhibitor is present in low quantities. The creation of enzyme inhibitors is thus a current area of biochemistry and pharmacology study (Knowles, 2014).

4.7.7. Enzyme Kinetics Modeling

The Michaelis-Menten equation and steady-state approximation are two mathematical models that may be used to explain and predict enzymatic processes. These models provide insight into the kinetics of enzymes, the mechanisms behind reactions and the factors influencing reaction rates. Enzyme kinetics researches how quickly they catalyze chemical reactions. In enzyme kinetics, the reaction's rate is established and the effects of altering its parameters are investigated. By analyzing an enzyme's kinetics in this way, it is possible to get knowledge about the catalytic mechanism of an enzyme, its role in metabolism, how its activity is controlled and how a drug or alteration (inhibitor or activator) might affect a rat. Typically, a protein molecule called an enzyme (E) helps another molecule called a substrate (S) react. This binds to the enzyme's active site to create an enzyme-substrate complex (ES). From there, the complex changes into an enzyme-product complex (EP) and from there, via a transition state (ES*), it turns into product P. The series of events that make up the mechanism (Knowles, 2014).

$$E + S \rightleftarrows ES \rightleftarrows ES \rightleftarrows EP \rightleftarrows E + P$$

In this example, it is assumed that the reaction has a single substrate and product. Examples of this include the enzyme triosephosphate isomerase, which catalyzes any reaction with a single substrate and a single result and mutases like phosphoglucomutase, which catalyze the transfer of a phosphate group from one location to another. These less common enzymes pale in comparison to those that catalyze two-substrate, two-product reactions, such as the NAD-dependent dehydrogenases like alcohol dehydrogenase that oxidize ethanol by NAD+. Reactions involving three or four substrates or products are feasible, albeit they are less common. Glyceraldehyde 3-phosphate dehydrogenase, for example, contains three substrates and two products, proving that there is no requirement that the number of products and substrates be equal (Srinivasan, 2021).

When enzymes like dihydrofolate reductase (shown to the right) bind a variety of substrates, enzyme kinetics may also illustrate the sequence in which the substrates bind and the products are released. An example of an enzyme that binds a single substrate and creates several products is a protease, which divides one protein substrate into two polypeptide molecules. Some link two substrates together, like DNA polymerase does when it joins a nucleotide to DNA. Despite the possibility of several complex phases in these systems, just one rate-determining step often determines the overall kinetics. This rate-setting procedure may involve an enzymatic reaction, a change in the enzyme's conformation, or substrates, such as those necessary for the enzyme to release the product(s) (Knowles, 2014).

Understanding kinetic data is aided by knowledge about the enzyme's structure. The structure, for instance, can reveal information about the changes that occur throughout the reaction, the binding of substrates and products during catalysis and even the purpose of certain amino acid residues in the mechanism. When an enzyme goes through a significant shape change throughout a process, it is helpful to know the structure of the enzyme both with and without bound substrate analogs. Many biological processes, such as RNA splicing and translation, rely on RNA-based catalysts like ribozymes and ribosomes. Biological catalysts are not only protein enzymes. The main difference between enzymes and ribozymes is that although RNA catalysts are composed of nucleotides, enzymes are composed of amino acids. Even though ribozymes only perform a limited number of activities, the same methods may be used to analyze and categorize their kinetics and processes. Enzyme activity offers information regarding flow, which is crucial to comprehending metabolic networks, in addition to metabolite or gene transcript data (Srinivasan, 2021).

Assumptions based only on, for example, gene transcript abundance alone may not be as accurate in this aspect since enzyme activity combines data from several levels of biological structure. The data sets produced by simultaneous measurements of, for example, the quantity of gene transcripts, the concentrations of metabolites and proteins and the activity of enzymes would be rich and likely include more information than the sum of their individual components. Traditional methods for measuring enzyme activity are labor-intensive and ineffective in comparison to the high throughput "omics" technologies presently being used to analyze the quantities of gene transcripts, proteins and metabolites. The platform for enzyme analysis that is described here is a step in the direction of the high throughput instrument that is typical in the "omics" discipline. However, because the biochemistry of measuring enzyme activity is far more complex than other high throughput approaches, true high throughput profiling on the size of genomes is unlikely. However, high throughput enzyme activity analysis has previously been developed (Srinivasan, 2021).

4.7.8. Enzyme Regulation

Allosteric control, covalent modification and enzyme induction/repression are a few methods for regulating an enzyme's activity. You may better comprehend how cells regulate enzymatic activities by understanding the kinetics of enzyme regulation. At least one of a metabolic pathway's enzymes—the regulatory enzyme—has either been activated, inhibited, or its concentration has altered when the rate of the pathway changes. Regulatory enzymes frequently catalyze the slowest or rate-limiting step in the pathway, therefore changing their rate will change the system's overall speed. The methods used to manage the rate-limiting enzyme in a process reflect the functioning of that pathway. Substrate Concentration. Every enzyme's rate is influenced by the substrate's concentration. As the concentration of the substrate [S] increases, the rate of enzymes increases to a maximum speed (Vmax), which occurs when the enzyme is completely saturable with the substrate. The Michaelis-Menten equation for many enzymes describes the relationship between vi (the initial reaction velocity), [S], Vmax and the Km (the substrate concentration at which vi = 12Vmax) (Knowles, 2014).

Some products are reversibly blocked by structural enzyme analogs and by-products. These inhibitors can be classified as competitive, noncompetitive, or uncompetitive according on how they impact the formation of the enzyme-substrate complex. The activity of an enzyme is regulated by substances known as allosteric activators or inhibitors that connect to locations other than the active catalytic site and change the shape of the enzyme, which affects the catalytic site. The activity of an enzyme can also be regulated by covalent changes, such as the phosphorylation of a serine, threonine, or tyrosine residue by a protein kinase. Enzyme activity can be regulated by the reversible binding of a modulator protein, such as Ca2+-calmodulin. Monomeric G-proteins (GTP-binding proteins) activate the targets proteins by reversible binding. Some enzymes are produced as zymogens, which are inactive precursors that are later activated by proteolysis. One example is the digestive enzyme chymotrypsin. The concentration of an enzyme can be regulated by alterations in the rate of enzyme synthesis (for instance, the activation of gene transcription) or the rate of degradation. A route's rate-limiting enzyme's regulatory mechanisms are always a reflection of how the process operates in a particular tissue. In feed-forward regulation, the substrate controls the process's speed; in feedback regulation, a pathway's final product either directly or indirectly controls how quickly it is synthesized. The biosynthetic and degradative processes are controlled by separate but coordinating regulatory systems. Pathways are also managed by compartmentalizing the enzymes unique to each route. An enzyme's activity can be regulated by other substances that either increase or decrease it. An enzyme's activity is decreased by inhibitors, whereas an enzyme's activity is increased by activators. Chemicals of various kinds can inhibit or increase the activity of enzymes in a number of ways (Knowles, 2014).

4.7.9. Competitive vs. Noncompetitive

Because an activator or inhibitor's binding is typically reversible, the molecule is not permanently bound to the enzyme. Some important groups of drugs have reversible inhibitory effects. Tipranavir, an HIV drug, is an example of a reversible inhibitor. Reversible inhibitors are divided into several types according to how they bind and one of these classes helps the virus multiply by inhibiting the activity of a viral enzyme. We won't go through every category, but we will concentrate on the two most important types: competitive inhibitors and noncompetitive inhibitors. When an inhibitor binds to an enzyme's active site, for example, it can prevent the substrate from attaching. Competitive inhibition is the process where the inhibitor "competes" with the enzyme's substrate. In other words, only the inhibitor or the substrate may be bound at any one time. The substrate is not inhibited from binding to the active site by the inhibitor in a noncompetitive inhibition. Instead, it attaches in a different place, obstructing the enzyme's ability to function as it should. The term "non-competitive" for this inhibition is due to the simultaneous binding of both the inhibitor and substrate (Srinivasan, 2021).

4.7.10 Enzyme Mechanism

Understanding the mechanism of enzyme catalysis is a key aspect of enzymology. Investigating the sequential actions that enzymes do to promote chemical processes, such as transition state formation, substrate binding and product release, is required for this. Enzyme mechanisms, together with the steps that make them up, are the terms used to describe the chemical alterations that are brought about by enzymatic action on substrates (Knowles, 2014).

Theoretically, the mechanism of catalysis by enzymes is similar to that of other types of chemical catalysis, such as when it combines several types of catalysis. While defining their operations, enzymes' kinetic and thermodynamic properties are established.

For more than a century, researchers have worked to understand the precise mechanism of enzyme catalysis and the components that give enzymes their astounding catalytic efficiency. It is known that enzymes may speed up processes by a ratio of more than 1017 times. A wealth of information about enzymes has been gleaned from experimental and computational research, particularly demonstrating the importance of enzyme structure in catalysis.

The induced-fit and transition-state stabilization theories, as well as the lock-and-key theory's original and updated variants, propose that enzymes serve as structural scaffolds that recognize and bind their substrates in certain ways. After binding and proper participant placement, it is recommended that enzymes provide a reaction environment that differs greatly from that of the bulk solvent and encourages the reaction to advance more quickly than is possible in solution.

It has been established that enzyme residues that make direct contacts with the substrate (and cofactor) are essential for maintaining the positions of the reaction participants and it is well recognized that mutations of these residues can alter the rate and/or outcome of the catalyzed process. It is well known that the active-site geometries correspond to the transition state of the catalyzed reaction and that the electrostatic environment of the active sites varies from that of the bulk solvent in a manner that facilitates reaction development. In terms of structural scaffolds, the function of enzyme active sites in supplying a complementary structure and electrostatic environment to preferentially bind and/or sustain the transition state has therefore long been acknowledged. Despite decades of research, many key aspects of enzyme catalysis remain unexplained. What roles in catalysis do the parts of the enzyme outside the active site play? Huge, intricate structures may be seen in enzymes. Little is known about the role of these residues in the catalytic mechanism, despite the fact that it is widely understood that residues located far from the active site can significantly alter how an enzyme catalyzes a chemical. There is proof that the kinetics and activity of enzymes can be considerably impacted by the hydration shell and bulk solvents. Only a fragmentary understanding exists of the interaction between enzyme structure and function and solvent. Furthermore, the catalytic efficiencies of bio-organic and even protein molecules designed intentionally to mimic the active site and have surroundings similar to the transition state of a catalyzed process are significantly lower than those of their naturally occurring counterparts (Knowles, 2014).

CHAPTER 5

Membranes and Surfaces in Biological Systems

Contents

A biological membrane, also known as a cell membrane or biomembrane, is a permeable membrane that either separates the interior of a cell from the outside environment or divides an intracellular region into several smaller compartments. The phospholipid bilayer present in eukaryotic cell membranes is in charge of aiding chemical and ion transport, as well as signal transmission. Because the concentration of lipids in a cell membrane is required for physiological activity, proteins in a fluid matrix may rotate and disperse laterally in all directions. Integral membrane proteins have an annular lipid envelope composed of lipid molecules that are tightly bound to their surface. This allows the integral membrane proteins to adapt to the lipid bilayer's high membrane fluidity environment. Mucous membranes, basement membranes and serous membranes, despite their names, are not cell membranes; rather, they are membranes that separate tissues made up of layers of cells (Chan, 2012).

5.1. COMPOSITION

5.1.1. Asymmetry of Biological Membrane

A lipid bilayer is made up of two parts: an exterior leaflet and an interior leaflet. The imbalance between the two sides is caused by the differential distribution of bilayer components.

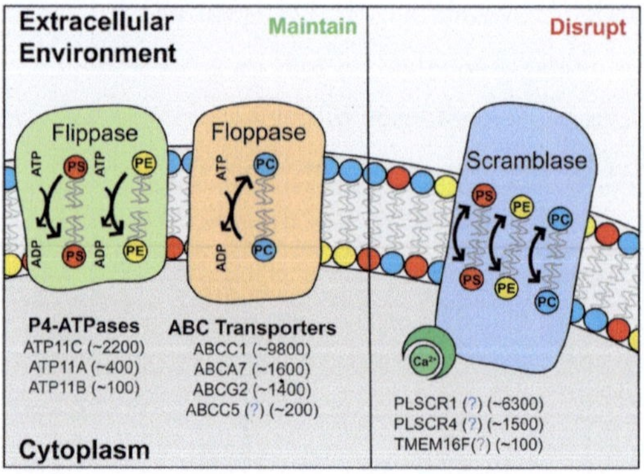

Figure 5.1. Membrane asymmetry in RBCs is regulated by flippase, floppase and scramblase enzymes.

Source: Fraser, Merryn & Matuschewski, Kai & Maier, Alex. (2021). Of membranes and malaria: phospholipid asymmetry in Plasmodium falciparum-infected red blood cells. Cellular and Molecular Life Sciences. Scientific Figure on ResearchGate. Available from: https://www.researchgate.net/figure/Membrane-asymmetry-in-RBCs-is-regulated-by-flippase-floppase-and-scramblase-enzymes_fig1_350042332 (accessed on 22 September 2023)

This imbalance promotes the efficiency of biological processes such as cell signaling. The biological membrane's asymmetry reflects the various roles performed by each leaflet of the membrane. The membrane's outer and inner leaflets do not fit together correctly. Certain proteins and lipids may be attached to just one side of a membrane (Figure 5.1) (Yan, 2014).

During membrane trafficking, the cytosolic and exoplasmic sides of the plasma membrane, as well as the proteins, lipids and glycoconjugates expressed on the extracellular side of the plasma membrane and facing the endoplasmic reticulum (ER) and the Golgi apparatus, are maintained. In eucaryotic cells, enzymes connected to the endoplasmic reticulum membrane and facing the cytoplasm are responsible for the creation of new phospholipids. These enzymes use free fatty acids as a substrate to transport freshly formed phospholipids to the cytosolic side of the bilayer. This is known as phospholipid translocation. To ensure that the membrane grows evenly, one-half of the newly formed phospholipid molecules must be transported to the opposing monolayer. Flippase enzymes assist to speed up the process. Certain kinds of lipids end up being concentrated in the different layers due to the selective nature of phospholipid transport across the plasma membrane through flippases. While selective flippases are useful, various approaches may be used to create asymmetric lipid bilayers. A distinct mechanism is at action when it comes to glycolipids, the lipids whose asymmetric distribution is most prominent and constant in animal cells (Chan, 2012).

5.1.2. Lipids

Lipids, which are components of the biological membrane, have hydrophilic heads and hydrophobic tails. Hydrocarbon tails, also known as hydrophobic tails, determine the size of the cell due to their length and saturation. The development of lipid rafts occurs when lipid species and proteins cluster together in membrane domains. These help to segregate membrane components into functional zones where activities like signal transduction may occur. Erythrocytes, often known as red blood cells, have a unique lipid content. Phospholipids and cholesterol are both components of red blood cell membranes. The erythrocyte membrane is important in the coagulation of blood. There is a chance that phosphatidylserine is present in RBC bilayers. In most situations, this occurs inside the cytoplasm, near to the cell membrane. It is subsequently delivered to the plasma membrane, where it may have a role in blood clotting (Yan, 2014).

5.1.3. Proteins

Phospholipid bilayers are densely packed with proteins. These membrane proteins serve as catalysts for a broad range of chemical processes and have a number of additional uses. Integral membrane-spanning proteins often contain several domains on both the outer and inner sides of the protein. Integral proteins are difficult to remove when there is a strong adhesion between lipid bilayers. They will not split into their component pieces until the membrane is chemically changed and torn. In contrast to integral proteins, peripheral proteins have relatively weak links to the bilayer's surface, making them easily dissociate. Integral proteins, on the other hand, have strong surface connections. Asymmetry on the opposite side of a membrane is caused by proteins that are only located on one side of the membrane (Chan, 2012). There are intrinsic membrane proteins. These proteins are present entirely or partially inside the plasma membrane, which envelopes the cell's core. Membrane proteins are proteins that are completely encased in their own membrane. The bulk of these proteins include hydrophobic side chains that interact with the fatty acyl group of membrane phospholipids to increase protein adhesion to the membrane. Almost all membrane-associated organisms include these proteins. Transmembrane proteins, such as integral membrane proteins, are examples of proteins that may have a single or several transmembrane passages. On both sides of the bilayer, these proteins contain domains that extend into the aqueous media. They also have one or more parts that span the membrane, which might be -helices or many -strands. Red blood cell glycophorins are single-pass transmembrane proteins, while RBC chloride-bicarbonate exchange proteins and band-3 proteins are multipass transmembrane proteins. Red blood cells contain glycophorins. Because of the phospholipid bilayer's endurance, separating transmembrane proteins using standard extraction procedures may be difficult (Chan, 2012).

There are also extrinsic membrane proteins. Because the membrane contains a hydrophobic core, membrane proteins cannot enter the cell from the outside. These proteins interact with other integral membrane proteins, which either tethers or transfers them to the membrane. The key variables that contribute to their capacity to stick to membranes are electrostatic and hydrogen bonding interactions. These peptides are sometimes referred to as proteins located in the peripheral membrane. Extrinsic membrane proteins, which are present in the area of the membrane closest to the cytoplasm, play a critical role in signal transmission. Other examples include the cytoskeletal proteins spectrin and actin, which may be found in red blood cells, as well as the enzyme Protein kinase C. Extrinsic membrane proteins may be effectively separated using solutions with a high pH, extremely high or low ionic strength and moderate extraction (Yan, 2014).

5.1.4. Oligosaccharides

As the third major component, the plasma membrane includes a substantial quantity of carbohydrates. After interacting with other molecules to form glycoproteins or glycolipids, they bind to the membrane's surface. These polysaccharide chains that are linked to membranes are known as proteoglycans. All eukaryotic cells have an outer membrane composed of carbohydrates with between 2 and 60 monosaccharide units that may be straight or branched. Glycoproteins are proteins that have at least one oligosaccharide attached to them. It is involved in two distinct biological processes: cell recognition and cell surface antigenicity. Glycolipids are a form of lipid that contains carbohydrates that are covalently bonded. They aid in the process of cell identification and play an important function in the preservation of cell integrity. Proteoglycans are proteins that exist in the extracellular matrix. Proteoglycans are lengthy polysaccharide chains that are covalently bound to the center of a protein. Simply defined, they are proteins that have a high glycosylation level. The molecule's glycosaminoglycan component is principally responsible for its major effect, which is that it allows tissues to retain water and expand in response to mechanical stress (Chan, 2012). Polymers such as oligosaccharides, which include sugar, are examples of polymers. By forming covalent connections with the lipids or proteins in the membrane, they may make glycolipids or glycoproteins. Both of these possibilities are plausible. Sugars are a component of the lipid molecules that comprise glycolipids, which are found in membranes. The sugar groups of glycolipids may be situated near the cell surface of the bilayer, where they have the capacity to establish hydrogen bonds. Glycolipids are the most extreme example of asymmetry in lipid bilayers. Glycolipids are a kind of lipid that may be found in biological membranes and serve numerous roles in communication, including cell identification and cell adhesion. Carbohydrates are the building blocks of proteins. They are essential important for the immune system to defend itself and react to potential threats (Yan, 2014).

5.2. MODELS OF MEMBRANES IN BIOLOGICAL SYSTEMS

We were ultimately able to construct the present idea and identify the properties of the fluid mosaic model after a thorough analysis of each suggested membrane model. The function of a few of the most notable membrane models produced by scientists will be discussed in the following sections.

5.2.1. Membrane-Based Theory Proposed by Gorter and Grendel (1920)

According to Evert Gorter and Francois Grendel's idea, if the plasma membrane is made up of two layers, the surface area of the lipid monolayer would seem to be considerably bigger than that of the plasma membrane. To put this theory to the test, blood cells from a variety of animals were employed and the area of the lipid monolayer as well as the overall surface area of the plasma membrane were measured using Langmuir's technique. They postulated that the plasma membrane is made up of two layers of lipids, with the hydrophilic head on the outside and the hydrophobic tail on the inside, with the hydrophilic head pointing away from the aqueous environment. The hydrophilic head would be on the exterior in this configuration. The predicted structure somewhat corroborated the researcher's theory. The model was unable to define the properties and functions of the membrane because inaccurate assumptions about the lipid population of the membrane were made (Chan, 2012).

5.2.2. The Sandwich Model, Created by Davson and Danielli in 1935

The cell membrane, according to Davson and Danielli, is made up of a lipid bilayer sandwiched between two extremely thin layers of protein on either side. They showed that the cell membrane has a lipoid core surrounded by protein monolayers. This core is located in the cell membrane's middle.

5.2.3. Robertson Devised the Unit Membrane Model in 1950

The paradigm devised by Robertson verified the sandwich model proposed by Davson and Danielli. He was the first to identify the trilaminar structure of the cell membrane using high-resolution electron microscopy. He reasoned that all biological membranes must have a common underlying structure, which he called the unit membrane and characterized as a lipid bilayer surrounded by protein sheets. Based on his findings, he came to this conclusion (Yan, 2014).

5.2.4. The Notion of Fluid Mosaic, Created by Singer and Nicolson and Originally Exhibited in 1972.

This specific variation has received the most attention up to this date. Proteins, according to this viewpoint, may be found safely encased inside the lipid bilayer that makes up the cell membrane. It seems that integral proteins, peripheral proteins and lipid-anchored proteins are dispersed throughout the membrane like tiles in a mosaic (Chan, 2012).

5.3. FORMATION OF MEMBRANES IN BIOLOGICAL SYSTEMS

A phospholipid bilayer is formed as a result of membrane lipids combining in water. The aggregation process is initiated by the hydrophobic effect, which happens when hydrophobic ends meet and get separated from water. The hydrogen bonds that exist between the hydrophilic heads and the water are strengthened as a result of this structure, while the repulsion that occurs between the hydrophobic tails and the water is decreased. When the entropy of a system increases, so does the amount of hydrogen bonds that are readily accessible. This starts a process on its own.

5.4. TYPES OF MEMBRANES IN BIOLOGICAL SYSTEMS

There are three types of membranes in biological systems. They include cell membrane, mucous membranes and S-layer.

5.4.1. Cell Membrane

The plasma membrane, also known as the plasmalemma or "phospholipid bilayer," is a semipermeable lipid bilayer that surrounds all cells. It links the cytoskeleton inside the cell to the cell wall outside the cell if it is present. Proteins and lipids, the two most prevalent biological components, perform a wide range of cellular tasks. In animal cells, the cell membrane encloses the cytoplasm and serves as a physical barrier between the cytoplasm and the outside environment, similar to the epidermis in plants (Figure 5.2). In contrast to vertebrates, where the cell wall serves as a selective boundary, the cell wall in fungi, bacteria and plants serves primarily as a mechanical support structure.

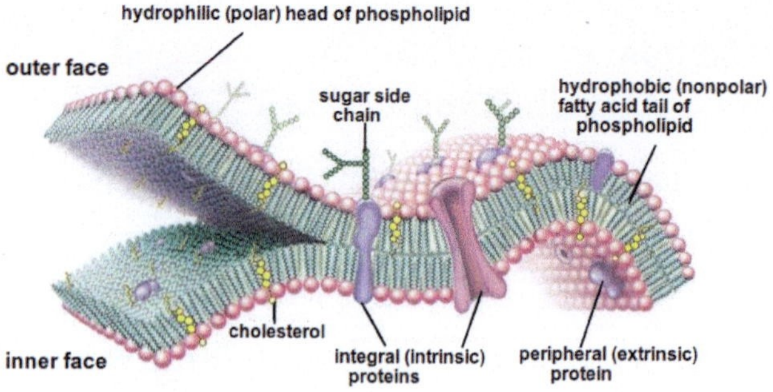

Figure 5.2. A simplified representation of a cell membrane.

Source: Ermilova, Inna. (2019). Modeling of biomembranes: from computational toxicology to simulations of neurodegenerative diseases. Scientific Figure on ResearchGate. Available from: https://www.researchgate.net/figure/A-simplified-representation-of-a-cell-membrane-Image-was-adapted-with-the-permission_fig1_333670261 (accessed on 22 September 2023)

Glues the cytoskeleton to the cell membrane, which aids in the organization of cells into tissues and gives them their distinctive form. Because it is selectively permeable and can control what enters and exits the cell, the barrier enables for the transfer of critical materials. There are two forms of material transfer across the membrane: passive transport, which requires no energy from the cell and active transport, which requires energy from the cell. The cell membrane also protects the functional capability of the cell (Chan, 2012).

Some proteins located inside membranes may act as molecular signals, allowing cells to communicate with one another. Receptor proteins are extensively distributed and capable of gathering information from their environment as well as other cells. These signals are received by the cell and converted into a format that allows it to reply rapidly. Additional proteins on the cell membrane serve as "markers" for other cells to recognize that particular cell. The interaction of these markers with their accompanying receptors is required for cell-cell communication in the immune system (Yan, 2014).

Structure

- *Lipidic bilayers*

The configuration of amphipathic lipid molecules in a lipid bilayer is seen schematically. The hydrophobic tails are isolated from the aqueous cytosolic and extracellular surroundings by yellow polar head groups. The cell membrane is composed of an extremely thin layer of amphipathic lipids that self-organize such that the hydrophilic "head" parts of the resultant bilayer interface with the cytosolic and extracellular sides, shielding the hydrophobic "tail" portions from the surrounding polar fluid.

A continuous, spherical lipid bilayer surrounds the cellular components, having a thickness of roughly 7 nm and being hardly observable under a transmission electron microscope. The cell may regulate the movement of hydrophilic solutes via transmembrane protein complexes like apertures and gates in the lipid bilayer by limiting passive diffusion of hydrophilic solutes through the band of hydrophobic tail groups. The enzymes Flippase and Scramblase concentrate phosphatidyl serine, a negatively charged lipid, on the inner membrane. This, along with NANA, inhibits charged Moities from permeating the membrane. Both bacterial and eukaryotic membranes serve significant roles. Managing what gets into and what comes out of cells is a critical task. Passive and active transport mechanisms, as well as membrane permeability, are explained by the phospholipid bilayer structure (fluid mosaic model) and specific membrane proteins.

Eukaryotic mitochondria and chloroplasts, like their prokaryotic counterparts, create ATP by chemiosmosis. Membrane proteins that are integral to the membrane. The proteins that make up the cell membrane are collectively referred to as integral membrane proteins. These structures, which may be seen via electron or fluorescence microscopy, can occur on either side of the membrane or even penetrate it. Some examples include synapses, desmosomes, clathrin-coated pits, caveolae and other adhesion-related structures (Chan, 2012).

- *The skeleton of the mesenchyme*

In the cytoplasm, the cytoskeleton forms projecting organelles that act as anchor sites for membrane proteins. Only some cell surfaces, such as the apical surface of epithelial cells that line vertebrate digestive tracts, allow anchoring proteins to permeate through the bilayer. The cytoskeleton is in charge of the formation of organelles that resemble appendages, such as cilia, which extend beyond the cell's surface but are still surrounded by the cell membrane. Microvilli are finger-like extensions on the apical surfaces of epithelial cells that increase cell surface area and speed up nutrient absorption. The cell membrane shields the cell from external dangers.

- *Fluid mosaic model and its constituent elements*

As postulated by S. Jonathan Singer and Garth Nicolson's fluid mosaic model, biological membranes may be conceived of as a two-dimensional liquid through which all lipid and protein molecules move relatively freely. If this image is true, it would be 10 nanometers in size. Synapses and desmosomes, as well as lipid rafts, pickets and fences, are examples of larger, more stable structures generated by the actin-based cytoskeleton in plasma membranes. The fluid mosaic concept is seen by fluorescently tagging the membrane proteins of two cells (say, human and mouse cells) with distinct colors. When the two cells are linked, the proteins are free to travel in the 2D plane, as shown by the mixing of the two colours (Chan, 2012).

Composition

Cell membranes include several biological components, including lipids and proteins. To deposit or remove chemicals from the membrane, many procedures are used:

When intracellular vesicles merge with the plasma membrane, their contents and membranes are recycled into the cell membrane. When blebs on the membrane are pinched off, vesicles may form. If the tube is attached to the membrane in a continuous method, the contents of the tube may be sucked in continually. Because stable membrane components are insoluble in water, their concentration in the aqueous phase is low; nonetheless, molecular exchange may still occur with this restricted reservoir. The mechanical strain in the membrane continually influences the exchange rate. Because of their elastic and dynamic mechanical properties, some cells with a generally uniform shape show a link between membrane tension and area. This dependency is known by several names, including homeostasis, zone control and stress management (Yan, 2014).

- *Lipids*

Phosphatidylcholine (PtdCho), phosphatidylethanolamine (PtdEtn), phosphatidylinositol (PtdIns) and phosphatidylserine (PtdSer) are examples of primary membrane phospholipids and glycolipids. The cell membrane is composed of amphipathic lipids such as phospholipids, glycolipids and steroids. Phospholipids are usually typically present in the highest concentrations in all cell types. In RBC tests, lipids were shown to make up around 30% of the plasma membrane. Both glycolipids and phospholipids contain fatty chains with 14 to 24 carbon atoms. Fatty acids with 16 or 18 carbons are the most frequent. There are both saturated and unsaturated fatty acids and the double bonds in fatty acids are often trans. The length of fatty acid chains and the degree of unsaturation have a significant impact on membrane fluidity. Despite the fact that the membrane's structure is relatively pliable and not fixed in place, it is held together by non-covalent interactions between hydrophobic tails. A cell membrane's phospholipid molecules are fluid in the sense that they may flow freely in all directions along the membrane's outermost layer. However, the interlayer mobility of phospholipid molecules is poor and hardly seen. Lipid rafts and caveolae are examples of cholesterol-rich membrane microdomains. Animal cell membranes often include varying quantities of cholesterol, which stiffens and strengthens the membrane by filling in the spaces between the hydrophobic tails of membrane lipids (Chan, 2012).

- *Carbohydrates*

Carbohydrates contribute for around 5% of the total mass of the plasma membrane, predominantly glycoproteins (cerebrosides and gangliosides), but also some lipoproteins. Unlike other cellular membranes, the plasma membrane is not extensively glycosylated. All cells contain a glycocalyx, but those with microvilli have the most. Recent research has indicated that the glycocalyx has a function in lymphocyte homing and cell adhesion. Galactose is the penultimate sugar and sialic acid is the terminal sugar during the transformation of the sugar backbone in the golgi apparatus. Due to its net negative charge, sialic acid may prevent positively charged particles from entering the body (Yan, 2014).

5.4.2. Mucous Membrane

Endodermal mucous membranes, commonly known as mucosae or mucosa, are bodily structures in charge of secretion and absorption of various chemicals. They coat the insides of many organs and cavities in the body that are exposed to the outer world. It does not separate from the epidermis in a few sites on the body, including the lips, nose, vaginal region and anus. Mucus is a viscous, sticky fluid that glands and mucous membranes release. It is in charge of keeping the body healthy and functional. Whether or whether the tissues in question create mucus, the phrase "mucous membrane" refers to the location of various organs and tissues inside the body. The bulk of the respiratory system is housed in the many cavities found throughout the human body. Mucous membranes make up the glans penis and clitoridis, as well as the interior of the prepuce (foreskin) and the clitoral hood. This is likewise true of the clitoridis (Figure 5.3).

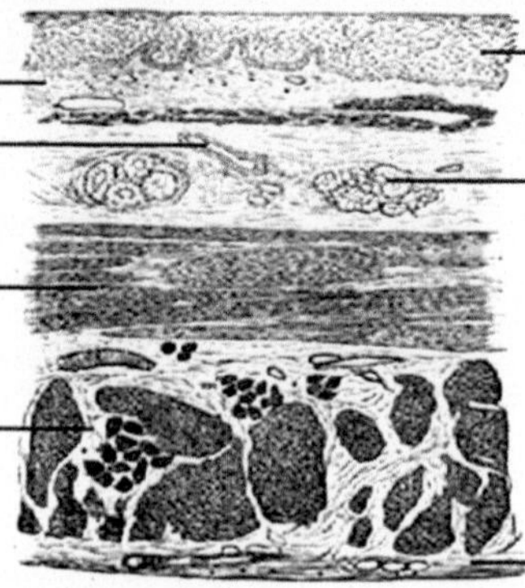

Figure 5.3. Mucous membrane of the esophagus.

Source: Saini, Sangita & Singh, Jasbir. (2015). Encyclopedia of Biomedical Polymers and Polymeric Biomaterials Mucoadhesive Polymers: Drug Delivery. Scientific Figure on ResearchGate. Available from: https://www.researchgate.net/figure/Mucous-membrane-of-the-esophagus_fig1_301816548 (accessed on 22 September 2023)

Connective tissue is a fibrous and elastic tissue that provides support to other physiological tissues and is often coated and protected by mucous membrane epithelial cells. Connective tissue aids in the preservation of the integrity of other physiological tissues (Chan, 2012).

Mucous Membranes of the Nasal, Pharyngeal and Eustachian Tubes

The existence of mucous membranes in parts of the body such as the nose, throat and ears is caused by the external environment (Yan, 2014) .

- ***Mucous membranes of the mouth***

Oral mucosa appears like lips, which are really membranes that surround the interior of the mouth and have a rosy pink color. These membranes are responsible for the lips' distinctive look. It is likely that chronic lip chapping is caused by inadequate hydration of the mucosal membranes. Saliva, in most situations, aids in the process of keeping the lips moist.

- ***Membranes lining the nasal chamber and sinuses***

Tiny blood vessels in your nasal mucous membranes have the ability to warm and moisten the air you breathe. Cilia, which resemble hair and cover their surfaces, act as a filter for dust inhaled and cover the surface of the cilia. The cilia then transfer the waste to the back of the neck or the front of the nose, depending on which position is most suited. A strong immune system is essential for preventing dangerous microorganisms from entering the body (Yan, 2014).

- ***The lining of the ear canal***

The middle ear is the body's first line of defense since it does not house any infections. They include cilia, which, like the mucous membranes of the nose, are responsible for pushing particles toward the ear canal. The presence of cilia in the mucous membranes that border the auditory canal helps foreign bodies migrate to the posterior of the pharynx, where they may be eaten.

Fluid accumulation in the middle ear may also be caused by middle ear diseases such as allergies and infections. The fluid may be infectious or sterile and it may or may not include white blood cells and IgA antibodies (Chan, 2012).

- ***The esophagus contains esophageal mucous membranes***

Peristalsis refers to the flow of food from the esophagus into the stomach, which is aided by the muscular component of the esophagus as well as the mucous membranes. Peristalsis is a digestive process that uses a wavelike rhythm to move food forward. Furthermore, the esophageal mucosal membranes contain several minute salivary glands that create a large quantity of bicarbonate. There is a chance that the bicarbonate will help to neutralize any stomach acid that has refluxed.

- ***The Mucous Membranes and Biological Aging Connection***

In contrast to the skin, which is continually exposed to the environment and may be harmed by UV radiation, mucous membranes within the body are rather protected. As a result, the mucous membranes are better equipped to preserve their structure as we age. Furthermore, the body's mucous membranes regenerate at a quick pace. However, studies have shown that as individual's age, their buccal mucosa thins. The Four Mucosal Membranes of the Oral Continent Reflect the Body. Because the mucous membranes of the mouth change in reaction to external stimuli, the oral cavity is frequently referred to as the "mirror of the body." Changes in symptoms are possible for the following illnesses:

Iron-deficiency Anemia is a kind of blood disorder, Sjogren's syndrome and other autoimmune diseases are quite real and Sarcoidosis is an infection of the lungs. Diabetes, Cushing's syndrome, hypothyroidism, hyperthyroidism and Addison's disease are all examples of disorders affecting the endocrine system. Infection brought on by Crohn's illness. Dehydration is one of the probable causes of dry mucous membranes, which may lead to a number of medical problems (Yan, 2014).

5.4.3. S-Layer

The S-layer is a component of the cell membranes of both archaea and bacteria. This layer is also known as the surface layer. Proteins or glycoproteins with structures similar to one another are layered in a monomolecular layer one on top of the other to create this layer. This framework, which creates itself in a flat, two-dimensional plane, covers the whole outside of the cell. As a consequence, it is possible that S-layer proteins account for between 10% and 15% of the total amount of proteins discovered in the cell [1, 2, 3]. Even in closely related species, S-layer proteins may change substantially and only a percentage of them are retained. This is because the S-layer has not been preserved in its entirety. [4]

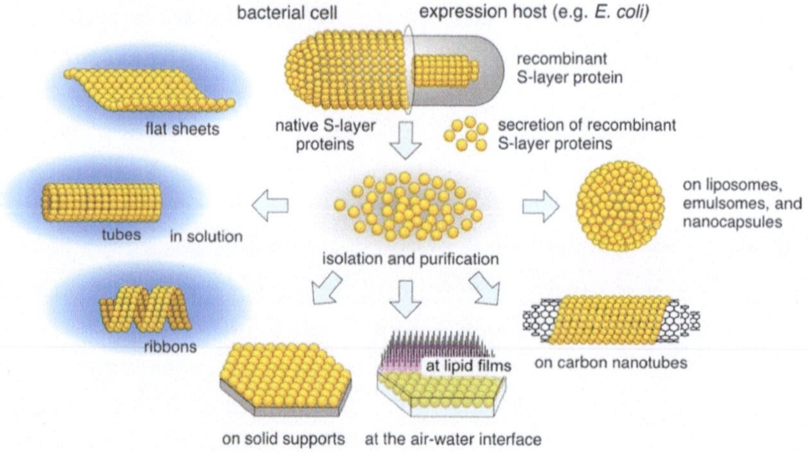

Figure 5.4. Schematic drawing of the different reassembly routes of S-layer proteins.

Source: Pum, Dietmar & Breitwieser andreas & Sleytr, Uwe. (2021). Patterns in Nature—S-Layer Lattices of Bacterial and Archaeal Cells. Crystals. Scientific Figure on ResearchGate. Available from: https://www.researchgate.net/figure/Schematic-drawing-of-the-different-reassembly-routes-of-S-layer-proteins_fig2_353504748 (accessed on 22 September 2023)

The thickness of the S-layers may range from 5 to 25 nm, while the hole size can range from 2 to 8 nm. S-layers may have lattice symmetry that is either square (p4), hexagonal (p3, p6), or oblique (p1, p2) depending on how they are built. At any one moment, the S-layer may include as few as one (P1) or as many as six (P6) identical protein subunits. This component of the structure is influenced by the lattice's symmetry. These components have unit cell widths ranging from 2.5 to 35 nanometers (nm), commonly known as center-to-center lengths (Figure 5.4) (Chan, 2012). The process through which the S-layer is attached to the cellular membrane can be well demonstrated. Gram-negative bacteria's S-layers connect to LPS by a number of methods, including ionic, carbohydrate-carbohydrate, protein-carbohydrate and/or protein-protein interactions. Gram-positive bacteria with an SLH domain may be able to attach themselves to peptidoglycan as well as a secondary cell wall polymer, such as teichuronic acids. If the SLH domain is present in the bacterial genome, this is the situation. The negatively charged polymer in the secondary cell wall acts as an electrostatic ligand for the S-layer protein's N-terminus. This is true even when SLH domains, which are also located in the secondary cell wall, are absent. Gram-negative archaea S-layer proteins contain a hydrophobic anchor in their structure that allows them to bind to the lipid bilayer. Methanochondroitin and pseudomurein bind to Gram-positive archaea S-layer proteins. These proteins are in charge of binding (Yan, 2014). The S-layer is the last barrier that separates a microbe from its surroundings and it is in charge of a number of critical duties. The S-layer is crucial for Gram-negative archaea's structural integrity since it is the only component of the cell wall. This is because it is the sole component of the cell wall. Furthermore, S-layers perform a variety of functions, including those mentioned below: phagocytosis, as well as resistance to certain kinds of bacteriophages. Low pH barriers are required to assure the safety of big molecules such as lytic enzymes. This defense might be due to glycosylated S-layers. The cytoplasmic membrane, peptidoglycan and exoprotein attachment sites are the three principal components responsible for preserving the integrity of Gram-positive prokaryotic membranes. They also carve out a home for themselves inside the periplasm (Chan, 2012).

5.5. FUNCTIONS OF MEMBRANES IN BIOLOGICAL SYSTEMS

Hydrophobic and hydrophilic properties may be observed in molecules with amphiphilic or amphipathic structures. Charged hydrophilic headgroups in phospholipid bilayers may form bonds with polar water. Furthermore, one layer's hydrophobic extensions interact with the adjacent layer's hydrophilic extensions. The length of the hydrophobic tails connected to fatty acids might vary. The fluidity of the lipid bilayer is regulated by lipid interactions, especially by the hydrophobic tails of the lipids. Cell membranes often create demarcated portions or compartments inside cells. These parts or compartments may sustain a continuous chemical or biological environment that is independent of the rest of the cell's environment. For example, the cell membrane separates a cell from its surroundings, while the peroxisomal membrane shields the rest of the cell from the potentially hazardous effects of peroxides. The waste products produced by cellular metabolism are kept in peroxisomes, which are vacuoles located within cells. Because the majority of organelles are identified by the membranes that surround them, they are classified as membrane-bound organelles (Yan, 2014).

A biomembrane's ability to selectively allow molecules to flow through it is likely to be its distinguishing feature. It should come as no surprise that atoms and molecules' chemical characteristics, such as size and charge, govern whether or not they can cross the barrier. Isolating a cell or organelle from its surroundings requires the capacity of the cell or organelle to selectively penetrate. Furthermore, due of their mechanical or elastic qualities, biological membranes may adapt to changes in shape as well as motion. Small hydrophobic substances may pass undetected through phospholipid bilayers (Chan, 2012).

Endocytosis, in which a vacuole attaches to a membrane and drives its contents into the cell, or membrane transport proteins are in charge of bringing in particles that are required for cellular activity but cannot freely diffuse across a membrane. The sarcolemma of muscle cells, neuron's specialized myelin and dendritic spine membranes and flagella, cilia, microvillus, filopodia and lamellipodia membranes are just a few examples of specialized plasma membranes that can isolate a cell from its external environment. Other examples include nerve cell sarcolemmae and dendritic spine and filopodia membranes. Caveolae, postsynaptic density, podosomes, invadopodiums, desmosomes, hemidesmosomes, focal adhesions and cell junctions are examples of "supramembrane" structures that plasma membranes may form. Invadopodiums are another example. Proteins and lipids are present in varying amounts within each kind of membrane. This category includes cell vesicles (phagosomes, autophagosomes and clathrin) as well as endosomes (rough and smooth endoplasmic reticulum, sarcoplasmic reticulum, Golgi apparatus, lysosomes, mitochondria (inner and outer membranes), nucleus (inner and outer membranes), peroxisome, vacuole and cytoplasmic granules). There are several types of biological membranes, each of which is made up of a unique combination of lipids and proteins. A membrane's composition may affect not only its physical but also its biological qualities. Membranes are made up of efflux pumps, which are responsible for the removal of chemicals from cells.

Because of the continual rotations that occur around the lipid tail links, the hydrophobic interior of the phospholipid bilayer is continuously shifting. Ends of hydrophobic bilayers flex and interlock with one another. The rotation and mobility of the hydrophilic head groups, on the other hand, are regulated by the hydrogen bonds formed with the water.

As the lipid bilayer approaches the hydrophilic heads, the viscosity of the lipid bilayer increases. When a lipid bilayer is chilled below the transition temperature, the highly mobile lipids begin to slow down and change into a solid with properties similar to gel. The temperature at which the transition occurs is influenced by the saturation of fatty acids inside the lipid bilayer as well as the length of the hydrocarbon chain. Fluidity that varies in reaction to temperature variations is a necessary physiological feature shared by all cold-blooded animals, including bacteria.

These organisms react to temperature fluctuations by changing the percentage of various fatty acids contained in the lipids that make up their membranes (Yan, 2014).

Cholesterol, a kind of sterol found in animal cells, is crucial for controlling membrane permeability. This molecule is prevalent in the plasma membrane, accounting for a large amount of the total lipid weight. Cholesterol molecules, which are tiny and stiff, may occupy the intermolecular gaps generated by bends in the unsaturated hydrocarbon tails of phospholipid molecules that are nearby. The bilayer has the potential to be reinforced by cholesterol, resulting in a more rigid and impenetrable bilayer. There are several reasons why the membranes of all cells must be fluid. It is vital in the process of cell signaling because it allows membrane proteins to interact with one another and rapidly diffuse throughout the bilayer plane. Diffusion allows membrane proteins and lipids to travel throughout the cell and into the bilayer. Diffusion starts at the sites of production. When a cell splits, the molecules that make up its membrane are rather evenly distributed throughout the daughter cells and the dividing membranes allow diverse substances to mix. It is difficult to imagine how cells might form and divide in the absence of fluid membranes, which are required for these activities.

The Helfrich model lays a strong focus on fluidity when calculating the amount of enthalpy required for elastic deformation of a membrane (Chan, 2012).

5.6. PROCESSES IN MEMBRANES OF BIOLOGICAL SYSTEMS

Membrane proteins are essential for both active and passive transport mechanisms, which transfer nutrients into and poisons out of cells. Transporters are a kind of protein that mediates the flow of solutes across membranes. Each member of this protein family attaches to a distinct cell type, which may be selected by differential transcription of the protein-coding genes. Transportation modes that need greater effort. Passive transport systems employ a variety of features, including as concentration gradients, hydrophobicity, size and charge, to move molecules without the usage of external energy. During this process, molecules shift from high concentration locations to low concentration areas (Yan, 2014).

The following are the three distinct passive transport processes:

When a barrier is only partly permeable to small molecules, they may freely diffuse along a concentration gradient. Fluids and gases such as water, carbon dioxide, ethanol and urea may flow through unhindered. Large molecules, such as ions and sucrose, may move and spread thanks to a protein called a transport protein. This kind of propagation is known as "facilitated diffusion." The circulation is employed in the process of delivering glucose and amino acids into the cell. When the concentration of a solute falls from a high to a low level, water molecules may flow across a selectively permeable membrane. The process of assimilation happens in the large intestine as nutrients in digested meals are transported osmotically from the stomach to the small intestine (Chan, 2012).

- ***Active transport***

Active transport systems carry molecules from a low-concentration region to a high-concentration area by using cellular energy. This is due to the fact that active transport mechanisms work against the concentration gradient. It prevents the concentrations of ions and molecules inside the cell from becoming imbalanced and fluctuating excessively.

- ***Active transport processes are classified into two types***

The majority of active transport across biological membranes is fueled by chemical energy in the form of ATP. The sodium-potassium pump is one of the most essential biological processes for managing the amounts of ions present within cells. This transport pathway uses ATP to propel potassium ions into the cell and sodium ions out of the cell. Another example of major active transport is the conversion of light energy into chemical energy in plant cells during photosynthesis. In order to create NADPH's reducing power, proteins must establish a photon gradient across the membrane. The energy stored in electrochemical gradients may be utilized in secondary active transport, which permits molecules to move in the opposite direction of concentration gradients. Given that ATP has nothing to do with the actual source of energy, the electrochemical potential difference created by ion pumping is the key determinant in determining secondary transport. The transport protein in a secondary active transport system is referred to as a symporter if it causes the transport of two molecules in the same direction and an antiporter if it causes the transport of two molecules in the opposite way (Yan, 2014).

Cystic fibrosis and other life-threatening disorders are caused by mutations in the CFTR gene. These mutations result in the development of faulty ion channels expressed by the gene. The cell's inability to export Cl-ions results in the formation of thick, dry mucus, which not only makes breathing difficult but also raises the risk of acquiring lung infections. Endocytosis is the mechanism through which tetanus neurotoxin and botulinum neurotoxin enter the cell. Once within the body, these neurotoxins disrupt neurotransmission and cause skeletal muscles to contract for a prolonged period of time. As a direct outcome of research into the structure and function of membrane-bound protein, effective novel drugs have been produced. A medicine containing pore-forming proteins has been created to treat influenza-related sickness (Chan, 2012).

- ***Movements of the membrane***

Despite the bilayer's ordered structure, the individual molecules that comprise it exhibit the random motion characteristic of liquids. Brownian motion is the name given to this kind of motion.

Phospholipid molecules in the bilayer have been demonstrated to migrate extremely swiftly in both directions. These motions, which occur 107 times per second and involve the locations of two molecules close to one another, may be monitored in seconds. Proteins located inside the biomembrane have varying degrees of mobility. Lipids, for example, are continually moving, but the proteins and lipids essential for the construction of membrane channels are rather stationary. As a direct result of this, membrane components move in a variety of ways. Molecules may move laterally inside the membrane, flip across the lipid layers and rotate completely within the membrane surface. The rotational diffusion coefficient may be used to quantify diffusion in a rotating medium (Yan, 2014).

The process of lateral diffusion occurs inside the membrane. This is frequently referred to as "floating" over the membrane sheet. Lipid molecules have free mobility inside their own layer within a two-dimensional liquid membrane and may flow in any direction. The last sort of motion is known as transverse diffusion and it is the least prevalent. They are quite uncommon in proteins. Biomembranes are distinguished from other membranes by an intrinsic asymmetry in both their structural and functional characteristics. This is reflected in the distribution of proteins as well as the makeup of the lipid's inner and outer layers. Several types of polar lipids have been added to both membranes to limit the number of times they flip back and forth. When lipids are introduced into biomembranes, ATP-dependent flippase enzymes cause them to flip (Chan, 2012).

5.7. SURFACES OF BIOLOGICAL SYSTEMS

A membrane cell is one example of a living entity that interacts with its surroundings via its surface. In natural systems, the major role of a membrane is to serve as a barrier to the flow of substances from the exterior into the interior, where they may subsequently engage in metabolic activities. It also serves as a mechanical stabilizer, a barrier, a chemical reservoir and a size and form divider, in addition to these roles. Natural membranes are very adept at detecting and selectively communicating just those targets while ignoring others. Supramolecular chemistry has traditionally depended on synthetic techniques that mimic natural membrane-based simple transporters for ions and small molecules by using either a carrier or channel mechanism.

Many manufactured carrier transporters are ionophores and siderophores that mimic their natural counterparts. The use of macrocyclic crown ethers as an ionophore for ion transport, for example, has been suggested. Crown ethers may assist the transfer of cations from aqueous solutions into nonpolar solvents due to their high affinity for alkali metals. They are useful in the process of phase transfer catalysis as a consequence of this characteristic (Yan, 2014).

A variety of options were investigated when developing artificial supramolecular biomimetics that work through channel-forming mechanisms. Biology drove the development of long helical polymers that mimicked the channels of the helix. However, since they lacked ion selectivity, membrane-spanning molecules could not self-assemble into channels using this technique. We studied the use of a size-based selection technique when developing macrocycle-based ion channel mimic chemicals. Macrocycles, which are amphiphilic chemicals, may be employed as selective ion filters to help ions flow through a membrane. By adding long substituents to a rigid ion-filtering molecule, chemical links may be formed between the spanning and size-selective components of an ion-filtering molecule. This allows the whole molecular assembly to be implanted in the membrane (Chan, 2012).

An example of an artificial channel-forming technique that uses helical molecules is the synthesis of membrane oligopeptides from serine and leucine utilizing peptides as the building blocks. Dimers and trimers of heptapeptides with different 21-residue sequences may be made. Because these dimers and trimers have a thickness of about 3–4 nm, they may permeate through phospholipid membranes. When the amino acids serine and leucine were combined, the resultant-helicates possessed a polar and hydrophobic surface. As a result, an interior cavity formed that was permeable to a wide range of alkali metals with minimal selectivity.

Size-selectivity approaches are used to create artificial ion channels from supramolecular macrocyclic building materials such as crown ethers, calixarenes and phthalocyanines. When it comes to transmembrane channels, for example, every member of the royal family has a significant affinity for a certain alkali metal. Ion channels based on the crown ether were successfully created using a potassium salt of tetracarboamido[18]-crown-6. Despite the fact that hydrated cations were scattered throughout the crown channels formed by the macrocycles, the crown ether substituents were not long enough to traverse a lipid bilayer. Cyclic tetraesters were used instead of anything else to provide the membrane-spanning function, with the crown functioning as a selectivity filter. This was done as a backup plan. Because of the low transfer rates for Li^+ generated by this approach, Na^+ and K^+ emerged as the more favored cations. Crown ethers were utilized as relays in a membrane-spanning design and their selectivity for potassium ions was shown to be 18 times greater than that of sodium ions (Chan, 2012).

Crown ether protocols, on the other hand, are not nearly as demanding. CDs and calixarenes, two cyclic compounds with clearly demarcated core compartments, have been incorporated in the formulations of these chemicals to improve the consistency and selectivity of the ion channel mimetic drugs. Rigid macrocycles were used instead of crown ethers, which are generally joined at the lipophilic section of the membrane. As a result, their hydrophobic substituents were able to cross through the lipophilic zone while remaining attached to the membrane's polar surface. Using this approach, we discovered that the selectivity of Na^+ in water was 68 times larger than that of K^+. Despite the phospholipid membrane's short length, this method proved effective in providing the needed selectivity for a synthetic ion channel (Yan, 2014).

CHAPTER 6

Protein Conformation

Contents

6.1. INTRODUCTION

Proteins are complex macromolecules that perform critical functions in living things. The precise three-dimensional shape, or conformation, of a protein, is required for its proper operation. The term "protein conformation" refers to the atomic arrangements and general structure that a protein has when it is unprocessed.The primary structure of a protein, which is the arrangement of amino acids in the protein chain, determines the protein's conformation. The hydrogen bonds, van der Waals forces and hydrophobic interactions that form between these amino acids influence the final conformation of the protein and drive the folding process (Morimoto, 2012). Protein conformation is dynamic; proteins can adapt to their surroundings, bind to ligands and interact with one another. These conformational changes frequently have a significant impact on the protein's function and regulation. Protein conformation is important in understanding how proteins perform the various roles they play in cells and organisms. It includes techniques such as X-ray crystallography, nuclear magnetic resonance (NMR) spectroscopy, and computational modeling that shed light on the three-dimensional structures of proteins and their dynamic behavior (Figure 6.1) (Wu, 2019).

6.2. PRIMARY STRUCTURE

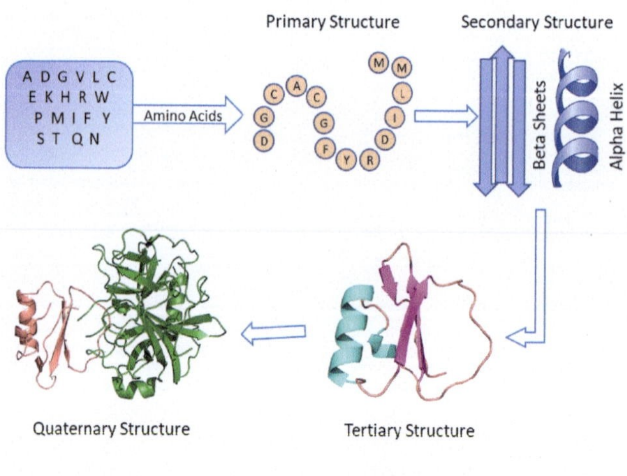

Figure 6.1. Illustration of types of proteins structure.

Source: Abbasi, Wajid Arshad & Minhas, Fayyaz ul Amir Afsar. (2019). Investigating Machine Learning Based Prediction of Protein Interactions. Scientific Figure on ResearchGate. Available from: https://www.researchgate.net/figure/

Different-levels-of-protein-structure-Different-amino-acids-joined-together-in-various_fig5_329208286 (accessed on 22 September 2023)

To build proteins, cells use 20 different types of standard L-amino acids. Amino acids, as the name implies, are made up of an acidic carboxyl group and a basic amino group. This dysfunctionality allows individual amino acids to join in long chains by forming peptide bonds—amide bonds between the -NH2 of one amino acid and the -COOH of another. Peptides are commonly used to describe sequences of 50 amino acids or less, whereas proteins and polypeptides are used to describe longer sequences. A protein can be made up of a polypeptide molecule or molecule. The carboxy-terminus, also referred to as the C-terminus, is the portion of a protein or peptide sequence that contains a free carboxyl group. The amino terminus and N-terminus of a sequence with a free-amino group are the termini of the sequence (Morimoto, 2012).

The amino acid's structural differences are due to the side chain substituents. Because of these side chains, the final peptide or protein has a variety of chemical, physical and structural properties. Figure 6.1 depicts the structures of the 20 most common amino acids found in proteins. Each amino acid has two abbreviations: one letter and three letters. These abbreviations are frequently used to make the written sequence of a peptide or protein easier to read. Depending on the side-chain substituent, an amino acid can be classified as acidic, basic, or neutral. We can only synthesize ten of the twenty amino acids required to produce the various proteins found in humans. The remaining ten amino acids, known as essential amino acids, must be obtained through diet (Wu, 2019).

The amino acid sequence of a protein is encoded in DNA. The two processes that go into the production of proteins are transcription, which involves using a DNA strand to create a complementary messenger RNA strand (mRNA) and translation, which uses the mRNA sequence as a template to direct the synthesis of the chain of amino acids that make up the protein. Post-translational modifications such as glycosylation or phosphorylation are frequently required for the protein's biological function. While a protein's primary structure is composed of its amino acid sequence, the three-dimensional or tertiary structure is critical to the protein's chemical and biological properties.

The primary structure of a protein is important because the order of amino acids affects how a protein folds and takes on its three-dimensional conformation. The folding process is propelled by interactions between amino acid residues along the polypeptide chain. Certain amino acid residues have a proclivity to form specific structural elements, such as alpha helices or beta sheets, based on their chemical properties and side-chain interactions.

The primary structure of a protein influences its stability, adaptability and functionality. Even minor changes in the amino acid sequence can have a significant impact on the conformation and function of a protein. For example, a single amino acid change can cause misfolding, which can lead to structural flaws and decreased protein activity (Morimoto, 2012).

The primary structure of a protein, or its amino acid sequence, serves as the starting point for structural analysis. This can be accomplished using both Edman degradation and protein sequencing based on mass spectrometry (MS). The former is particularly useful for generating complete protein sequences containing more than 40 amino acid residues. De novo protein sequencing is now possible thanks to the advancement of MS-based proteomics, which allows for the rapid and accurate determination of the sequence of any given protein. Peptide mapping, another MS-based technique useful for confirming sequence and identifying sequence variations such as point mutations, can be used to investigate a protein's primary structure. Understanding the primary structure is critical when studying protein conformation because it is the foundation for understanding how proteins fold and function. It is a starting point for research into secondary, tertiary and quaternary protein structures, among others. Researchers can interpret protein folding and conformational dynamics by predicting potential structural motifs and interactions using primary structure analysis (Wu, 2019).

6.3. SECONDARY STRUCTURE

Protein or peptide stretches or strands have distinct, recognizable secondary structural conformations that rely on hydrogen bonding. The two primary types of secondary structures are the ß-sheet and the -helix. The -helix's right-handed coil is coiled. The side-chain substituents of the amino acid groups extend outside of a -helix. Each C=O bond in the strand and each N-H group in the helix (four amino acids below it) form hydrogen bonds. This structure is particularly stable due to the hydrogen bonds. The side-chain substitutes of amino acids fit next to the N-H groups.

Intra-strand hydrogen bonds do not form in a ß-sheet; instead, inter-strand hydrogen bonds form. Strand pairs are arranged in a sheet shape and laid side by side. The adjacent strand's amino hydrogens form a bond with one of the strand's carbonyl oxygens. The two strands can be parallel or anti-parallel depending on whether the strand directions (N-terminus to C-terminus) are the same or opposite. The more aligned the hydrogen bonds, the more stable the anti-parallel ß-sheet.

Secondary structure refers to the regional folding patterns in a protein. It is largely determined by hydrogen bonding interactions between nearby amino acid residues along the polypeptide chain. The two most common types of secondary structures are beta sheets and alpha helices (Morimoto, 2012).

Alpha helices are formed when a polypeptide chain twists into a right-handed helical structure. The carbonyl oxygen of each peptide bond forms a hydrogen bond with the amide hydrogen of the amino acid four residues upstream to form an alpha helix. This predictable hydrogen bonding pattern keeps the helix structure intact. The side chains of amino acids extend beyond the helix axis. Alpha helices are frequently found in areas of the protein that must be stable, such as structural cores or transmembrane regions (Wu, 2019).

In contrast, beta sheets form when adjacent polypeptide chain segments line up side by side to form a sheet-like structure. A beta-sheet is formed when the amide hydrogen of one peptide bond meets the carbonyl oxygen of another peptide bond, either in the same or a different polypeptide chain. Beta sheets can be parallel or antiparallel depending on how the peptide strands are oriented. The side chains of the amino acids alternately extend above and below the sheet. Beta sheets are frequently found in protein structures involved in protein-protein interactions and the formation of protein cores.

Secondary structure components that are important for protein conformation and function include beta sheets and alpha helices. They participate in protein-protein interactions, provide structural stability to the protein and help shape it overall. Furthermore, the precise arrangement of secondary structure elements frequently defines functional sites within the protein, such as regions that bind ligands or active sites for enzymatic activity (Morimoto, 2012).

The most common technique for examining a protein's secondary structure is Circular Dichroism spectroscopy (CD). It measures various secondary structural components in a protein, such as a helix, sheet and random coil, based on their distinct spectra in the far-UV region of the spectrum. These spectra can then be used to calculate the fraction of each secondary structural component in the protein. A different biophysical technique known as Fourier transform infrared (FTIR) spectroscopy can be used to estimate secondary structural elements in a protein by measuring the wavelength and intensity of infrared radiation (IR) absorption by a protein sample (Wu, 2019).

Experiments such as X-ray crystallography and NMR spectroscopy are frequently used to determine the presence and arrangement of secondary structure elements within a protein. Understanding the secondary structure's organization and content allows researchers to gain insight into the protein's folding pattern, stability and potential functional implications.

6.4. TERTIARY STRUCTURE

The term "tertiary structure" refers to the three-dimensional arrangement of a protein's entire polypeptide chain (Figure 6.2). It represents the overall folding pattern that gives a protein its unique and distinctive conformation as a result of interactions between distant amino acid residues. The tertiary structure is required for the protein to function properly (Morimoto, 2012).

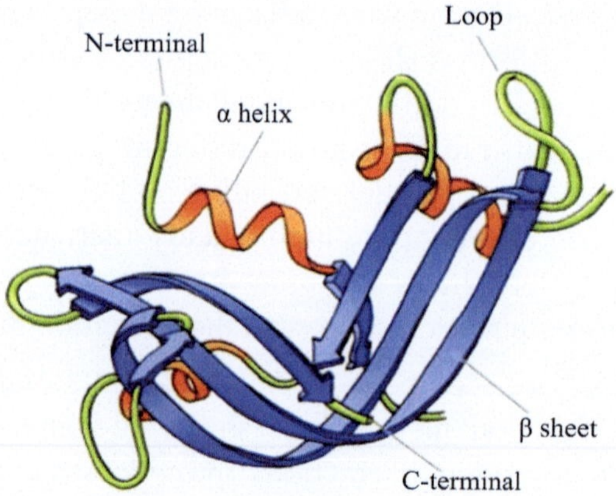

Figure 6.2. Tertiary structure of a protein.

Source: Biswas, Sunanda. (2021). CHAPTER-1 PROTEIN. Scientific Figure on ResearchGate. Available from: https://www.researchgate.net/figure/Schematic-representation-of-the-Tertiary-Structure-of-a-protein-Tertiary-structure-of-a_fig2_348370508 (accessed on 22 September 2023)

The folding of a protein into its tertiary structure is governed by a complex interplay of various forces and interactions. Van der Waals forces, hydrophobic interactions, disulfide bridges, hydrogen bonds and electrostatic interactions are examples of these. The backbone atoms of the polypeptide chain form hydrogen bonds with one another, which aid in the stabilization of specific structural components of the protein. Between cysteine residues, disulfide bridges form, which covalently link cysteine residues and increase protein stability.

Hydrophobic interactions are extremely beneficial to protein folding. Nonpolar amino acids tend to group in the interior of the protein away from water molecules, whereas polar or charged amino acids are more exposed to the solvent. As a result of this hydrophobic effect, the protein folds to maximize nonpolar interactions in the core while minimizing exposure of hydrophobic residues to the surrounding water (Wu, 2019).

The tertiary structure of a protein determines its overall shape, which is required for its function. Enzyme active sites, for example, have specific shapes that allow them to bind substrates and catalyze chemical reactions. Binding sites on transport proteins, like receptors, allow them to recognize and interact with specific molecules. The precise positioning of alpha helices and beta sheets in the secondary structure also influences the overall fold and stability of the protein. In addition to its function, a protein's tertiary structure can affect its stability and susceptibility to denaturation. Denaturation is the loss of function caused by a disruption in the natural structure of a protein. Protein unfolding can result in denaturation due to factors such as temperature, pH and the presence of denaturing agents (Morimoto, 2012).

The main methods for determining a protein's tertiary structure, or three-dimensional structure, are Cryogenic Electron Microscopy (cryo-EM), Nuclear Magnetic Resonance (NMR) spectroscopy and X-ray crystallography. Each method has advantages and disadvantages. These techniques typically involve collecting various experimental data types (such as chemical shift and X-ray diffraction) to measure atom distances and produce an atomic-level three-dimensional model of the protein structure. As of this writing, the 3D structures of over 190,000 different proteins had been determined. Finally, hydrogen-deuterium exchange (HDX)-MS is a powerful analytical method that is frequently combined with other structural methods to obtain critical data on protein conformation, dynamics and folding, yielding a more complete picture (Wu, 2019).

Understanding the tertiary structure of proteins is critical for many types of research and applications. Experiments such as X-ray crystallography,

NMR spectroscopy and cryo-EM can be used to determine the three-dimensional structure of proteins to learn more about their functions and mechanisms. Protein tertiary structures are predicted and examined using molecular modeling and simulations, as well as other computational techniques. Understanding protein tertiary structure is critical in areas such as drug design, protein engineering and understanding the molecular causes of diseases.

6.5. QUATERNARY STRUCTURE

Many proteins are composed of numerous polypeptide chains, also known as protein subunits.

A heterodimer and a homodimer have different subunits, while the reverse is true for a homodimer. The quaternary structure refers to the interaction and arrangement of these protein subunits to form a larger aggregate protein complex. Once again, a variety of interactions such as hydrogen bonds, disulfide bridges and salt bridges help to stabilize the final shape of the protein complex.

Quaternary structure refers to the arrangement of multiple protein subunits or polypeptide chains to form a useful protein complex. In other words, it describes how individual protein subunits interact and combine to form a larger, biologically active assembly (Morimoto, 2012).

Non-covalent interactions used to join two or more subunits together in quaternary structure proteins include hydrogen bonds, electrostatic interactions, van der Waals forces and hydrophobic interactions. These interactions keep the complex stable and allow it to perform specific tasks that a single subunit cannot handle on its own.

The subunits of a protein complex can be homomeric or heteromeric and their arrangement can change to produce different quaternary structures. Interaction interfaces between subunits frequently involve complementary shapes and chemical properties, allowing for precise binding and recognition.

The quaternary structure of proteins has a significant impact on their stability, functionality and regulation.

It enables the creation of complex molecular devices such as transporters, receptors and enzymes, which rely on numerous subunits to function properly. The interactions between the subunits in the quaternary structure can affect the complex's enzymatic activity, ligand binding affinity and allosteric regulation (Wu, 2019).

Understanding quaternary structure is critical for fully understanding protein mechanisms and functions. It sheds light on the complex cellular processes involving protein-protein interactions and other molecules. Quaternary structure is also useful in drug discovery because it allows for the targeting of specific protein-protein interfaces.

Experiments such as X-ray crystallography, NMR spectroscopy and cryo-electron microscopy (cryo-EM) are frequently used to determine the quaternary structure of protein complexes. Scientists can see how the subunits are arranged and investigate their interactions using these techniques (Morimoto, 2012).

6.6. STABILITY OF PROTEINS

Protein stability is defined as the energy difference between a protein's folded and unfolded states in solution (Figure 6.3). Surprisingly, the free energy difference between these states is typically between 20 and 80 kJ/mol, which corresponds to one to four hydrogen bonds. Although proteins are only marginally stable, their stability is sufficient to prevent spontaneous unfolding at standard temperatures.

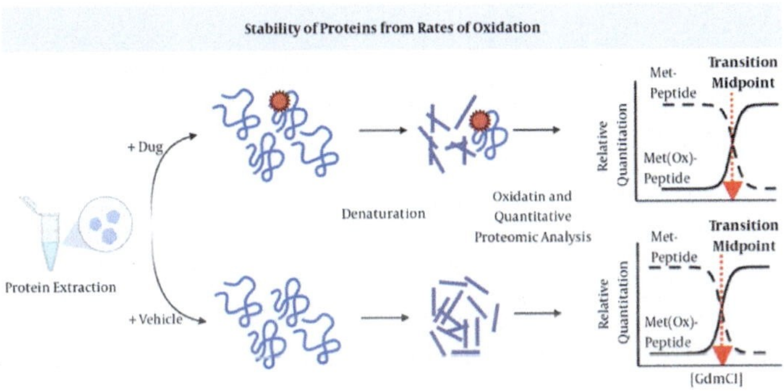

Figure 6.3. Stability of proteins from rates of oxidation.

Source: Gholizadeh, Elham & Rezaei tavirani, Mostafa & Emadi, Alireza & Karbalaei, Reza & Khaleghian, Ali. (2021). A New Drug Discovery Approach Based on Thermal Proteome Profiling to Develop More Effective Drugs. Middle East Journal of Rehabilitation and Health Studies. Scientific Figure on ResearchGate. Available from: https://www.researchgate.net/figure/Stability-of-proteins-from-rates-of-oxidation-SPROX-38-The-drug-target-interaction_fig2_351474963 (accessed on 22 September 2023)

When a protein folds, a large number of weak interactions can be observed. When the protein is unfolded, these interactions must be balanced against an almost identical number of interactions with water. Because each amino acid residue has the potential for polar interactions via the peptide bond as well as a variety of ionic, polar and nonpolar interactions via its side chains, this is a difficult problem to solve. This explains why determining a protein's structure solely based on its amino acid sequence is difficult because any energy computation errors are much larger than the protein's net stability (Wu, 2019).

- **The hydrophobicity effect**

The main driving force behind protein folding is the hydrophobic effect. Hydrophobic molecules are those that prefer to avoid contact with water. As a result, the hydrophobic side chains are buried inside the protein during protein folding. The behavior of hydrophobic molecules in water is complicated physically, but it is best understood in terms of their thermodynamic properties. The thermodynamics of protein folding, it turns out, closely resemble the behavior of simple hydrophobic molecules in water. Much of what we know about the hydrophobic effect comes from studies on the transfer of hydrocarbons from the liquid phase into water (Morimoto, 2012).

Hydrocarbon model studies show that entropic rather than enthalpic factors dominate hydrophobic compound insolubility at room temperature. When a hydrophobic molecule is introduced into an aqueous environment, a diffusely ordered water shell forms around it, similar to how clathrates form around noble gases and simple hydrocarbons. This is commonly referred to as the water-ordering effect. A shell forms around the hydrophobic compound due to its inability to form hydrogen bonds with the water surrounding it. As a result, the water molecules in question have fewer neighbors with whom to fulfill their hydrogen bonding potential. Entropy is reduced by lowering their degrees of rotational freedom. Because the standard changes in entropy and enthalpy for the transfer of a hydrophobic molecule to water are temperature dependent, this simple explanation based on water order is insufficient to fully explain the behavior of hydrophobic molecules in solution. Surprisingly, the value for the free energy change for the transfer of a hydrocarbon to water is temperature insensitive due to compensatory changes in entropy and enthalpy. These temperature dependencies are caused by the large temperature-insensitive heat capacity of hydrophobic molecules in solution. The water-ordering effect appears to be responsible for the abnormally high heat capacity of hydrophobic compounds in water.

While the molecular mechanisms underlying the hydrophobic effect are debated, there is a strong relationship between an organic molecule's free energy of transfer to water and its hydrophobic surface area. In contrast, there is no agreement on the relationships between solute molecular properties and entropic or enthalpic changes. Numerous studies have shown that the transfer of 1 A2 of the hydrophobic area to water results in an unfavorable increase in free energy of 80–100 J/mol. Given that this correlation is discovered to be largely independent of the characteristics of the hydrophobic solute, it is abundantly clear that the hydrophobic effect is a fundamental property of water (Wu, 2019).

According to studies on the heat capacity changes seen at the protein folding transition, protein denaturation is comparable to the conversion of hydrophobic molecules to hydrophilic molecules. Furthermore, it is well understood that the difference in exposed hydrophobic surface area between unfolded and folded states directly correlates with protein stability. Site-directed mutagenesis recently demonstrated the same thermodynamic relationship between the hydrophobic buried surface area and stability as seen for the transfer of organic hydrocarbons into water. That is, each buried 1 A2 of the hydrophobic surface area contributes 80 J/mol to the protein's stability when the only difference is the change in surface area. A protein behaves differently in solution than a simple hydrocarbon. Proteins, as previously stated, are only marginally stable. The decrease in entropy of the polypeptide chain as it adopts a well-defined conformation appears to be slightly offset by the change in the exposed hydrophobic surface area caused by protein folding. This explains why predicting the structure of proteins is so difficult (Morimoto, 2012).

- **Hydrogen bonds (Figure 6.4)**

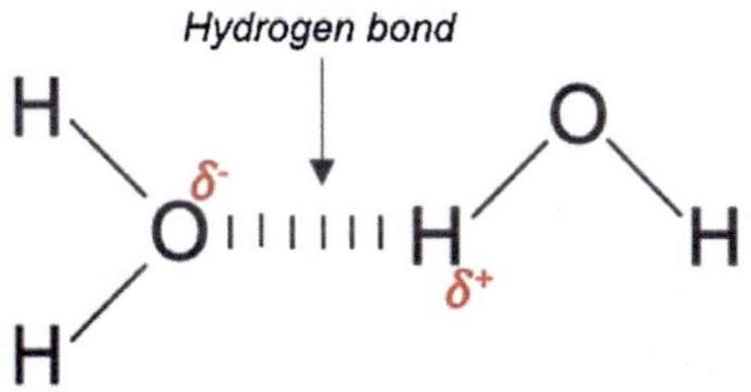

Figure 6.4. Diagram of hydrogen bonding.

Source: Christensen, Greg & Lou, Ding & Hong, Haiping & Peterson, G.P.. (2021). Improved Thermal Conductivity of Fluids and Composites Using Boron

Nitride (BN) Nanoparticles through Hydrogen Bonding. Thermochimica Acta. Scientific Figure on ResearchGate. Available from: https://www.researchgate.net/figure/Diagram-of-hydrogen-bonding_fig2_350425225 (accessed on 22 September 2023)

A hydrogen atom attached to an electronegative acceptor atom (D-H) and another electronegative acceptor atom (A) carrying a single pair of electrons interacts to form electrostatic hydrogen bonds (D-H—-A). In biological systems, the electronegative atoms are typically oxygen or nitrogen. When the donor and acceptor atoms are separated by 2.8 to 3.1. A, the D-H bond tends to be collinear with the lone pair of electrons. The geometry of the hydrogen bond is somewhat variable, which is consistent with the interaction's electrostatic dominance. For example, the N—H bond in the a-helix and antiparallel j-sheet is roughly colinear with the C=O bond rather than aligned with the lone pairs of the carbonyl oxygen. Many hydrogen bonds in proteins are formed in networks of interactions between acceptors and donors, with each acceptor interacting with multiple donors. This is consistent with the fact that the hydrogen bonds in proteins are ionic (Wu, 2019). It was initially thought that hydrogen bonds played a significant role in protein stability because of the extensive hydrogen bonding seen in a-helices and j-sheets. Almost every hydrogen donor and acceptor in a protein interacts with either the folded structure or an external solvent. The difference in free energy between a protein's folded and unfolded states, on the other hand, determines its stability. In the unfolded state, the polar components can form hydrogen bonds to water that are comparable to those found in the protein's tertiary structure. Hydrogen bonding is thus energetically neutral in terms of protein stability, with the caveat that any hydrogen bonding deficits in a folded protein are thermodynamically very unfavorable (Morimoto, 2012).

Although they do not affect stability, hydrogen bonds play an important role in protein conformation. The necessity of forming hydrogen bonds can be attributed to the abundance of a-helices and j-strands in protein structures.

- ***Disulfide bonds***

Many extracellular proteins contain disulfide bonds. Disulfide bonds greatly increase the stability of these proteins' folded states, as in many cases, simply reducing the cystine linkages is enough to cause unfolding. The source of the stability appears to be entropic rather than enthalpic. The addition of a disulfide bond reduces the entropy of the unfolded state by limiting the degrees of freedom that the disordered polypeptide chain can exercise. This stabilizes the folded state by reducing the entropy difference between the folded and unfolded states. This suggests a simple method for introducing disulfide bonds into proteins to improve protein stability. Although it may appear to be a simple task, the geometry of the disulfide bond is quite constrained. As a result, there are very few places in a protein where two residues can be replaced with cysteines without jeopardizing the protein's ability to fold under heat. T4 lysozyme studies have shown that, if appropriate sites can be found, the size of the closed loop produced by forming a disulfide bond is proportional to the degree of protein stability introduced.

- ***Ionic exchanges***

The majority of proteins have a salt bridge, also known as an ion pair, or the association of two ionic groups with opposing charges within the protein. Because the isolated ionic groups are so well solvated by water, these interactions have little effect on protein stability. As a result, there are very few unsolved salt bridges in the interior of proteins. Furthermore, salt bridges are rarely conserved in orthologous proteins (Wu, 2019).

- ***Dipoles interactions***

Dipole-dipole interactions are weak interactions caused by the close association of induced or permanent dipoles. Van der Waals interactions are the collective name for these forces. These interactions are common in proteins and vary in strength (Morimoto, 2012).

The strongest interactions, which are critical to the peptide bond, are observed between permanent dipoles. The alignment of the dipoles connected to the carbonyl and amide groups in the peptide creates a significant dipole moment in the peptide bond (3.5. Debye units versus 1.85 for a water molecule). These interactions decay with the inverse of the second to third power when the dipoles are fixed and with the sixth power when they are free to rotate. As a result, the amino-terminal end of an a-helix, where the dipoles are constrained and aligned, has a significant positive dipole. As a result, in enzyme active sites, negatively charged ligands are frequently bound to the N-terminus of an a-helix.

Permanent dipoles can also generate a dipole moment in a nearby atom or group. It is, however, far weaker than the stabilizing interaction observed between permanent dipoles. These interactions are important because they change the relative charges of the atoms, which has a significant impact on the activation barriers of enzyme-catalyzed reactions.

The London or dispersion forces of a dipole-dipole are the weakest. These can be understood qualitatively as the result of a temporary asymmetry in a neutral atom's charge distribution, which induces a favorable dipole in a nearby neutral atom, resulting in a weak attraction. The best way to describe them is through quantum mechanics. These forces have an inverse relationship with the sixth power of separation. Even though these forces are very weak, a folded protein contains a large number of them, which contributes significantly to the folded state's stability. The Van der Waals forces are important in stabilizing interactions between proteins and their complementary ligands, whether the ligands are proteins or small molecules (Wu, 2019).

6.7. ANALYSIS OF PROTEIN STRUCTURE

Due to the complexity of protein structure, elucidating a complete protein structure is extremely difficult even with the most sophisticated analytical tools. An amino acid analyzer can tell you which amino acids are present and their molar ratios. Peptide mapping, Edman degradation, or mass spectroscopy can then be used to examine the protein sequence. This process is simple for peptides and small proteins, but it becomes more difficult for large multimeric proteins.

During peptide mapping, different protease enzymes are typically used to treat the protein to break up the sequence into smaller peptides at specific cleavage sites. Trypsin and chymotrypsin are two commonly used enzymes. Mass spectroscopy has evolved into an important tool for the analysis of enzyme-digested proteins due to the use of peptide fingerprinting techniques and database searching. Edman degradation begins at the N-terminus of a short peptide and involves the cleavage, separation and identification of one amino acid at a time (Morimoto, 2012).

Circular dichroism spectroscopy (CD) is one technique used to determine the secondary structure of a protein. In the far-UV region of the spectrum (190–250 nm), the circular dichroism spectra of the various secondary structure types, including ß-sheet, random coil and -helix, are all distinct. The percentage of each type of structure in the entire protein that these spectra can estimate.

To perform a more thorough, high-resolution analysis of a protein's three-dimensional structure, X-ray crystallography or nuclear magnetic resonance (NMR) analysis is used. To analyze the three-dimensional structure of a protein using X-ray diffraction, a large, well-ordered single crystal is required. The three-dimensional electron density map produced by X-ray diffraction, which also allows measurement of the small distances between atoms, can be used to create a model of the protein structure. NMR has some advantages over X-ray diffraction for determining the three-dimensional structure of proteins because it can be used in solution and frees the protein from the constraints of the crystal lattice. NOESY and COESY are the most commonly used two-dimensional NMR techniques for measuring atom separations through space and bonds, respectively (Wu, 2019).

A protein's stability can be determined using a variety of techniques. To study the unfolding of a protein, spectroscopic techniques such as fluorescence, UV, Infrared and CD can be used. One of the thermodynamic techniques, Differential Scanning Calorimetry (DSC), can be useful in determining how temperature affects protein stability. Comparative peptide mapping, which is frequently performed using LC/MS, is a very useful method for identifying chemical alterations in a protein, such as oxidation or deamidation. HPLC is a useful tool for determining the purity of a protein. Protein stability can also be determined using additional analytical techniques such as SDS-PAGE, isoelectric focusing and capillary electrophoresis and the potency of a protein biopharmaceutical should be determined using a suitable bioassay. Arrayed instruments are now widely available for tracking "particle" size over time and under different conditions to determine the state of aggregation (Morimoto, 2012).

6.8. CONFORMATIONAL CHANGES

Protein conformational changes are changes in the three-dimensional structure of a protein that causes it to change from one conformation to another. These modifications are required for protein regulation, interaction and operation with other molecules. Protein conformational changes can be localized, affecting only a few amino acids, or large-scale, affecting the entire protein structure.

Because proteins are naturally flexible, they can adopt a variety of conformations. Many factors, including ligand binding, pH or temperature changes, post-translational modifications and interactions with other proteins or molecules, can cause conformational changes. Proteins must make these adjustments, which frequently involve switching between active and inactive states or allowing interactions with other molecules, to carry out their biological functions.

Proteins undergo several conformational changes, including:

- Folding is the process by which a protein transitions from an unstructured or partially structured state to its native, three-dimensional conformation. Because of this conformational change, the protein can assume its functional state. When a protein loses its native structure and transitions to a denatured or unfolded state, this is known as unfolding. Temperature, pH and the presence of denaturing agents are just a few of the many variables that can influence the dynamic folding and unfolding processes.
- Domain Movements: Proteins are frequently made up of distinct structural domains, which are small, self-contained units of protein folding. Domain movement refers to the repositioning of these domains in relation to one another. These conformational changes can be critical for protein function, as seen in enzymes that move many of their domains to accommodate catalysis or substrate binding. Domain movements are also possible in response to allosteric regulation, in which the binding of a ligand at one site results in conformational changes at a distal site (Wu, 2019).
- Loop or Segment Motions: Proteins have flexible regions, such as loops or segments, that can change their conformation. While maintaining the overall fold of the protein, these regions can adopt a variety of conformations. Loop movements are critical in protein-ligand interactions, substrate binding and catalysis. Large-scale movements involving flexible segments, such as hinge bending or twisting, can influence how proteins function or aid in protein interaction.
- Helix-Coil Transitions: Some proteins have alpha-helical regions that can switch between helical and unstructured coil conformations. Temperature changes or binding activities are frequently associated with these helix-coil transitions. Helix-coil transitions, for example, allow DNA-binding proteins to switch between being bound to the DNA and having helical conformations and being free of the DNA and having coil conformations (Morimoto, 2012).
- Allostery: Allostery is the process by which a signal is transferred from one site in a protein to another, causing the protein's conformation to change and its function to change. This conformational change could be caused by environmental factors

or the binding of a ligand. Allosteric conformational changes are required for protein-protein interactions, signal transduction and enzyme activity regulation. They frequently involve long-distance communication between different protein regions through changes in interactions and dynamics.

- Oligomerization: Numerous proteins can oligomerize and assemble to form multimeric complexes, in which different protein subunits combine to form a functional unit. Conformational changes are frequently involved when these complexes are assembled or disassembled. Specific ligands or substrates, for example, can bind to one another and cause conformational changes that promote complex formation. Conformational changes, on the other hand, may cause subunits to separate or protein complexes to disassemble.
- Aggregation and Amyloid Formation: Under certain conditions, proteins can undergo conformational changes that result in abnormal protein aggregation. Insoluble protein aggregates such as amyloid fibrils, which have been linked to several neurodegenerative diseases, may form as a result of these conformational changes. The transition from a native to an aggregated conformation involves protein misfolding and the adoption of beta-sheet-rich conformations.

The study of protein conformational changes necessitates the use of both experimental and computational methods. High-resolution images of various protein conformations can be obtained using methods such as cryo-electron microscopy, NMR spectroscopy and X-ray crystallography. Computational techniques, such as molecular dynamics simulations, can be used to simulate protein conformational dynamics, which can also shed light on the energetics and pathways of conformational changes (Wu, 2019).

Understanding the mechanisms of protein function, protein-protein interactions and regulatory processes necessitates a thorough understanding of protein conformational changes. Because conformational changes can be used to modulate protein activity or prevent disease-related aberrant protein conformations, they have implications for drug discovery and design.

6.9. PROTEIN AGGREGATION AND MISFOLDING

Proteins, which are intricately folded molecules, rely on their precise three-dimensional structures to perform biological functions. Proteins, on the other hand, can misfold under certain conditions, which is defined as adopting incorrect conformations that diverge from their natural structures. Protein misfolding can lead to aggregates, which are soluble protein clumps or fibrils that can have serious consequences. These diseases, which are linked to this process, are known as conformational or protein misfolding diseases (Morimoto, 2012).

Protein misfolding can be caused by a variety of factors, including genetic mutations, environmental stressors, cellular environment adjustments, or errors in the protein folding machinery. When proteins misfold, hydrophobic regions that are normally hidden within the core of the native structure of proteins become visible. These exposed hydrophobic patches can interact with other misfolded proteins or partially unfolded intermediates and they are prone to self-association.

Misfolded proteins can aggregate in a variety of ways, depending on the protein and the context of the disease. One common type of protein aggregation is the formation of soluble oligomers, which are compact collections of improperly folded proteins. These oligomers can then combine to form larger aggregates like amyloid fibrils and amorphous aggregates. Amyloid fibrils are insoluble protein structures with a distinct cross-beta sheet conformation that makes them stable and resistant to degradation.

Cellular toxicity can occur as a result of the accumulation of protein aggregates, which can disrupt cellular functions. These aggregates can disrupt protein transport, interfere with cellular organelle function and trigger a chain reaction of cellular reactions like oxidative stress and inflammation. Protein aggregates, which can act as seeds or templates for the misfolding and aggregation of additional proteins, can also produce a self-sustaining cycle of protein misfolding and aggregation (Wu, 2019).

Many neurodegenerative diseases, including Alzheimer's, Parkinson's, Huntington's and Prion diseases, are strongly linked to protein misfolding and aggregation. During these disorders, certain proteins, such as amyloid-beta, tau, alpha-synuclein, huntingtin, or prion proteins, misfold and assemble in the affected tissues, resulting in the formation of toxic protein deposits. These aggregates have the potential to impair neuronal function, harm cells and eventually contribute to neuronal degeneration and cognitive decline.

The mechanisms underlying protein misfolding and aggregation are important areas of research in the field of protein conformation. The causes of protein misfolding, the mechanisms that lead to aggregate formation and the toxic effects of protein aggregates are all being investigated by researchers. With this knowledge, strategies to prevent or reduce protein misfolding and aggregation may be developed, potentially leading to the development of treatments for diseases caused by abnormal protein conformation (Morimoto, 2012).

6.10. EXPERIMENTATION METHODS FOR INVESTIGATING PROTEIN CONFORMATION

Understanding structure-function relationships and elucidating the biological functions of proteins rely heavily on conformational research. Numerous experimental methods have been developed to study protein conformation at various levels of detail. These techniques provide a window into a protein's dynamic behavior as well as valuable insights into the three-dimensional arrangement of its atoms. A few commonly used experimental methods for examining protein conformation are as follows:

- X-ray crystallography is a popular method for determining the atomic structure of proteins. In this process, a protein sample is crystallized and information about the protein crystal's X-ray diffraction is gathered. By analyzing the diffraction patterns, the positions of the atoms within the protein can be determined by reconstructing an electron density map. X-ray crystallography can determine high-resolution protein structures, allowing for a thorough analysis of protein conformation (Wu, 2019).
- Nuclear magnetic resonance (NMR) spectroscopy is an effective method for studying the conformation and dynamics of proteins in solution. NMR exposes the protein sample to radiofrequency pulses and a strong magnetic field, causing the atomic nuclei to resonate at specific frequencies. These resonance frequencies can be measured to learn valuable information about dynamics, local and global structure. NMR is especially useful when studying flexible and disordered protein regions because it reveals conformational changes.
- Cryo-Electron Microscopy (Cryo-EM): Cryo-EM has revolutionized structural biology by allowing the visualization of protein structures, including large protein complexes, at nearly atomic resolution. By flash-freezing proteins in vitreous ice, cryo-EM preserves their native conformation. An electron microscope is then used to image the samples and a three-dimensional

reconstruction of the protein is created by combining multiple images. Cryo-EM analysis of protein conformation in a state close to its native state can provide insights into dynamic protein assemblies and their conformations.

- Small-Angle X-ray Scattering (SAXS): SAXS is a solution-based method, is used to investigate the general structure and conformation of proteins in solution. This requires exposing a protein sample to X-rays and measuring the scattering pattern of the X-rays as they interact with the protein molecules. The scattering data can be used to determine the protein's size, shape and overall conformation. SAXS is useful for studying flexible proteins and protein complexes in solution (Morimoto, 2012).
- Spectroscopy of Circular Dichroism (CD): CD spectroscopy investigates how chiral molecules, such as proteins, absorb left- and right-circularly polarized light differently. This method reveals information about secondary structure and protein conformational changes. Because CD spectroscopy is relatively fast, it can be used to track conformational changes caused by temperature, pH, ligand binding, or denaturing agents in real-time.
- Fluorescent Spectroscopy: Fluorescent spectroscopy makes use of the light emitted and absorbed by fluorescent molecules. Fluorescent probes can be used in protein conformation studies to target specific areas of a protein to track changes in fluorescence properties such as intensity, polarization, or lifetime, which can be a sign of conformational changes. Fluorescence spectroscopy, due to its versatility, can shed light on protein folding, stability, ligand binding and conformational dynamics.
- Hydrogen-deuterium exchange mass spectrometry (HDX-MS): HDX-MS is a powerful technique for studying protein conformational dynamics and interactions with other proteins or ligands. Deuterium atoms in a deuterated solvent are exchanged for labile hydrogen atoms in a protein. The degree of deuterium incorporation into various protein regions is determined using mass spectrometry. HDX-MS allows for the analysis of conformational changes caused by elements such as ligand binding or protein-protein interactions, as well as information about the solvent accessibility and stability of various regions of a protein (Wu, 2019).

We've learned a lot about protein conformation and how it relates to protein function, folding and dynamics by combining experimental and computational methods. By providing detailed structural information, these methods pave the way for advancements in a variety of fields, including drug discovery, enzymology and structural biology.

Methods based on computers

Computational methods are important in the study of protein conformation because they reveal information about structural dynamics, folding mechanisms and protein conformational changes. Computer simulations and mathematical algorithms are used in these techniques to model and predict protein structures, simulate their movements and examine conformational changes at atomic resolution.

Molecular dynamics (MD) simulations are a popular computational method for studying protein conformation. Using MD simulations to model the movement of atoms in a protein over time, researchers can observe the dynamic behavior and protein conformational changes of proteins. By resolving the equations of motion for each atom per classical physics rules, MD simulations can reveal information about protein flexibility, folding pathways and interactions with other molecules. MD simulations also allow for the investigation of unusual conformational events as well as the computation of kinetic and thermodynamic properties (Morimoto, 2012).

Another computational method used in protein conformation studies is homology modeling, also known as comparative modeling. Homology modeling attempts to predict the three-dimensional structure of a protein using experimentally determined structures of related proteins. By aligning the target protein sequence with known structures, computational algorithms can generate a model of the target protein's conformation, including side chain positions and overall fold. When using experimental structure determination techniques such as X-ray crystallography or NMR spectroscopy, homology modeling can be especially useful (Wu, 2019).

In addition to MD simulations and homology modeling, other computational techniques are used to study protein conformation. Algorithms for minimizing system energy, which optimizes the atomic positions of proteins to produce stable conformations, are among them. Energy minimization is frequently used to improve protein structures before MD simulations or in conjunction with other techniques. Exploring protein conformational landscapes and calculating free energies are also accomplished through the use of sophisticated methods such as enhanced sampling methods. Enhanced sampling techniques such as meta-dynamics, replica exchange MD, or accelerated molecular dynamics overcome energy barriers to sample rare conformations and transitions between different states. Calculations of free energy, such as thermodynamic integration or umbrella sampling, provide quantitative information on the populations and stability of various conformations.

Computational approaches to protein conformation also include observational data analysis and structural feature prediction. Docking algorithms, for example, can predict a protein's conformation and binding mode in a complex with a ligand. Protein tertiary structure can be predicted using only the protein's sequence data using techniques such as ab initio folding and threading.

Computational methods for protein conformation, in general, provide useful tools for understanding how dynamic proteins are, investigating conformational changes and forecasting protein structures. They aid in the search for new drugs and direct the development of novel proteins with desired properties.

In conclusion, conformation is a critical component of protein structure and function. The three-dimensional configuration of a protein's atoms governs its stability, activity and interactions with other molecules. Because of the dynamic nature of protein conformation, proteins can undergo conformational changes that are required for their various functions (Morimoto, 2012).

Computational methods have completely changed the way protein conformation is studied by shedding light on protein dynamics, folding mechanisms and conformational changes. Researchers have been able to investigate protein structures and forecast conformational states using methods such as homology modeling, energy minimization and molecular dynamics simulations. These computational techniques have significantly improved our understanding of protein behavior and have aided in the development of new medications and therapeutics (Wu, 2019).

Experiments such as X-ray crystallography, NMR spectroscopy and cryo-EM are still extremely useful for determining protein structures. Computational approaches, on the other hand, supplement these experimental methods by providing atomistic details, forecasting conformational changes and providing insights into the thermodynamics and kinetics of protein conformation. Protein conformation research has far-reaching implications in fields such as biochemistry, molecular biology and drug discovery. Understanding the relationship between protein structure and function allows us to decipher the mechanisms underlying biological processes, develop more potent drugs and engineer proteins with desired properties (Morimoto, 2012).

CHAPTER 7

Mass Spectrometry

Contents

7.1. INTRODUCTION

Mass spectrometry is a powerful analytical technique that allows the identification, quantification and structural analysis of molecules based on their mass-to-charge ratio. It has become an indispensable tool in various scientific disciplines, including chemistry, biochemistry, pharmaceuticals, forensics and environmental science. This section provides an overview of mass spectrometry and its historical development and highlights its importance in scientific research (Kliem, 2010).

Mass spectrometry is a technique that involves the ionization of molecules followed by their separation based on their mass-to-charge ratio and subsequent detection (Figure 7.1). It enables the determination of a sample's elemental or isotopic composition, identification of unknown compounds, determination of molecular structure and quantitative analysis (Makarov, 2013).

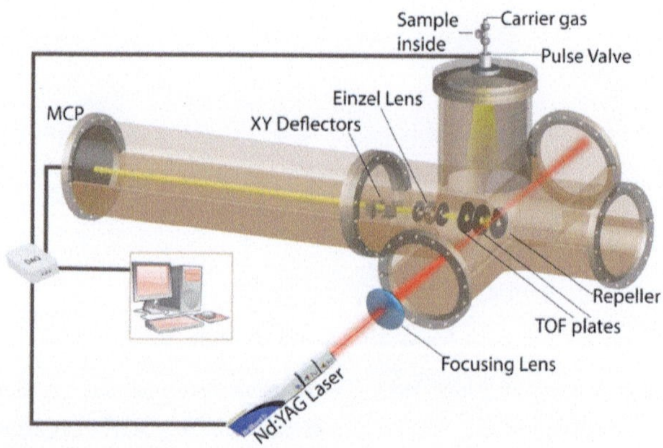

Figure 7.1. Schematic of the time-of-flight mass spectrometer setup developed for multiphoton ionization and dissociation experiments.

Source: Chacko, Roby & Barik, Saroj & Banhatti, Shreyak & G. Aravind. (2022). Multiphoton ionization and dissociation of polycyclic aromatic hydrocarbon molecules of astrophysical interest. Physical Review. Scientific Figure on ResearchGate. Available from: https://www.researchgate.net/figure/Schematic-of-the-time-of-flight-mass-spectrometer-setup-developed-for-multiphoton_fig1_359157826 (accessed on 22 September 2023)

The process of mass spectrometry involves several essential steps. Firstly, the sample is ionized by different ionization methods such as electron impact ionization (EI), chemical ionization (CI), electrospray ionization (ESI), matrix-assisted laser desorption/ionization (MALDI), or atmospheric pressure chemical ionization (APCI). These methods generate ions by removing or adding electrons to the molecules, forming charged species. The ions are then introduced into the mass analyzer, which separates them based on their mass-to-charge ratio.

Different mass analyzers are used, including magnetic sector analyzers, quadrupole mass analyzers, time-of-flight (TOF) mass analyzers, ion trap mass analyzers and Fourier transform ion cyclotron resonance (FT-ICR) mass analyzers. Each mass analyzer has unique advantages and limitations, determining the instrument's resolution, mass accuracy and mass range (Kliem, 2010).

Following the separation of ions, they are detected by ion detectors, such as electron multipliers, Faraday cups, *channeltron* detectors, or microchannel plate detectors. These detectors convert the ions into electrical signals, which are then processed by data acquisition systems, including analogue-to-digital converters, time-to-digital converters and data processing software. The resulting data can be analyzed to obtain valuable information about the sample (Makarov, 2013).

The development of mass spectrometry dates back to the early 20th century when British physicist J.J. Thomson first observed the deflection of charged particles in a magnetic field, leading to the invention of the first mass spectrometer, the parabola spectrograph. Since then, mass spectrometry has undergone significant advancements, both in terms of instrumentation and methodologies. One of the groundbreaking developments in mass spectrometry came with the introduction of the quadrupole mass analyzer in the 1950s. This compact and versatile analyzer revolutionized the field by offering high sensitivity, mass range and resolution. It quickly became the preferred choice for many applications.

In the 1970s, the introduction of electron impact ionization (EI) and chemical ionization (CI) techniques expanded the capabilities of mass spectrometry. EI provided a reliable and reproducible method for ionizing volatile compounds, while CI allowed for thermally labile and non-volatile compounds to be analyzed.

The advent of soft ionization techniques in the 1980s, such as electrospray ionization (ESI) and matrix-assisted laser desorption/ionization (MALDI), opened up new possibilities for analyzing large biomolecules, such as proteins and nucleic acids. These techniques allowed for ionising delicate biomolecules without extensive fragmentation, preserving their structural information (Kliem, 2010). In the 21st century, we witnessed significant advancements in mass spectrometry, including the development of high-resolution mass spectrometers capable of achieving mass accuracies in the parts-per-million range. This advancement has enabled the identification and characterization of complex mixtures with unprecedented accuracy and precision. Mass spectrometry plays a crucial role in scientific research across various disciplines. Its importance stems from its ability to provide detailed and precise information about molecules' composition, structure and behavior (Makarov, 2013). In the field of chemistry, mass spectrometry is widely used for the identification and quantification of organic compounds. It is essential in drug discovery and development, environmental analysis and studying complex chemical reactions. Mass spectrometry enables researchers to determine the elemental composition of compounds, identify impurities and study reaction mechanisms. Mass spectrometry is instrumental in analyzing proteins, peptides and other biomolecules in biochemistry and proteomics. It allows for the identification and sequencing of proteins, post-translational modification analysis and protein-protein interaction studies. Mass spectrometry is also used in metabolomics, enabling the comprehensive analysis of small molecules involved in metabolic pathways. In pharmaceutical research, mass spectrometry is used for pharmacokinetic studies, which involve the measurement of drug concentrations in biological samples. Mass spectrometry can quantify drugs and their metabolites in complex matrices, providing valuable information on absorption, distribution, metabolism and excretion (Kliem, 2010). Forensic science heavily relies on mass spectrometry to analyze drugs, explosives, toxins and other substances of forensic interest. It plays a critical role in drug testing, crime scene investigation and the detection of illicit substances. Mass spectrometry also contributes to environmental science by enabling the identification and quantification of pollutants in air, water and soil samples. It aids in monitoring environmental contamination, assessing the impact of human activities on ecosystems and supporting ecological regulations. Mass spectrometry is a versatile and powerful analytical technique with many applications. Its importance in scientific research is evident in its contributions to chemistry, biochemistry, pharmaceuticals, forensics and environmental science. By providing valuable information about the composition, structure and behavior of molecules, mass spectrometry continues to drive advancements and discoveries in numerous fields (Makarov, 2013).

7.2. FUNDAMENTALS OF MASS SPECTROMETRY

Mass spectrometry is based on several fundamental principles that enable the analysis of molecules based on their mass-to-charge ratio. This section explores mass spectrometry principles, including ionization methods, mass analyzers, ion detection and data acquisition systems.

7.2.1. Principles of Mass Spectrometry

Mass spectrometry principles involve the ionization of molecules, their separation based on mass-to-charge ratio and subsequent detection. These principles are fundamental to the functioning of mass spectrometers (Kliem, 2010).

Ionization methods

Ionization methods are crucial in mass spectrometry by converting neutral molecules into charged ions. Various ionization methods are employed depending on the nature of the analyte and the desired ionization efficiency.

- *Electron Impact Ionization (EI)*

Electron impact ionization (EI) is a widely used ionization method in mass spectrometry. The analyte molecules are bombarded with high-energy electrons during EI, typically at 70 eV. This energetic bombardment causes the ejection of an electron from the molecule, forming a positively charged molecular ion (M+). Additionally, the molecular ion may undergo fragmentation, leading to the generation of fragment ions. The resulting fragment ions provide valuable structural information about the analyte molecule, aiding in its identification and characterization within the mass spectrometer (Figure 7.2).

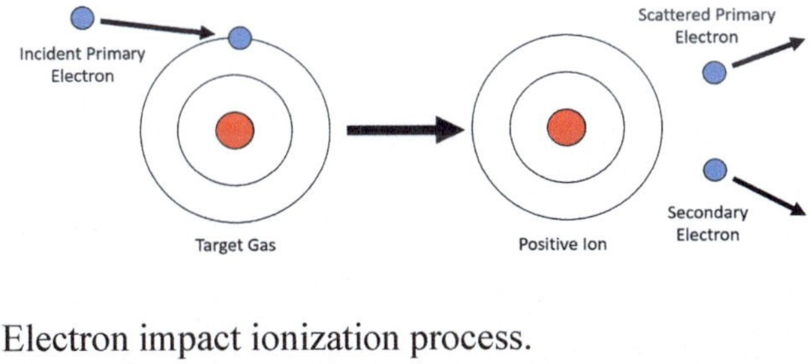

Figure 7.2. Electron impact ionization process.

Source: Chacko, Roby & Barik, Saroj & Banhatti, Shreyak & G., Aravind. (2022). Multiphoton ionization and dissociation of polycyclic aromatic hy-

drocarbon molecules of astrophysical interest. Physical Review A. Scientific Figure on ResearchGate. Available from: https://www.researchgate.net/figure/Schematic-of-the-time-of-flight-mass-spectrometer-setup-developed-for-multi-photon_fig1_359157826 (accessed on 22 September 2023)

- *Chemical ionization (CI)*

Chemical ionization (CI) is an ionization method employed in mass spectrometry that operates at higher pressures than electron impact ionization (Figure 7.3). In CI, the analyte molecules are subjected to reactions with ions of a reagent gas present in the ion source. These reactions lead to the formation of protonated or ionized molecules. CI is particularly suitable for analyzing thermally labile and non-volatile compounds unlike electron impact ionization. The gentle ionization conditions of CI reduce the chances of sample fragmentation, allowing for detecting and characterizing intact molecular ions and providing valuable information about the analyte's structure and composition (Makarov, 2013).

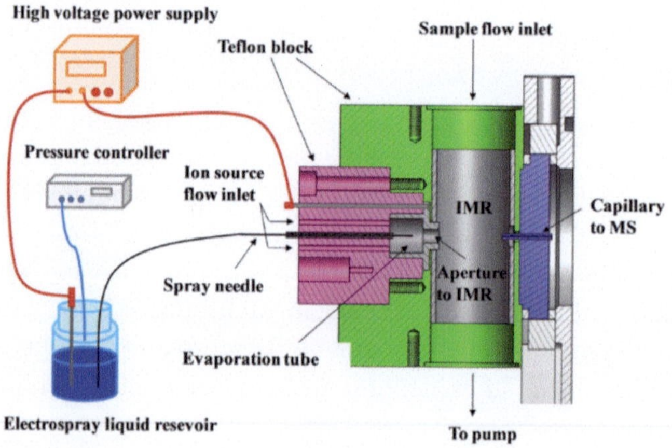

Figure 7.3. Chemical ionization mechanism.

Source: Zhao, Yue & Chan, Jeremy & Lopez-Hilfiker, Felipe & McKeown, Megan & D & apos;Ambro, Emma & Slowik, Jay & Riffell, Jeffrey & Thornton, Joel. (2017). An electrospray chemical ionization source for real-time measurement of atmospheric organic and inorganic vapors. Atmospheric Measurement Techniques. Scientific Figure on ResearchGate. Available from: https://www.researchgate.net/figure/Schematic-of-the-electrospray-chemical-ionization-ESCI-source-module-Also-shown-are_fig1_320214632 (accessed on 22 September 2023)

- *Electrospray ionization (ESI)*

Electrospray ionization (ESI) is a highly versatile ionization technique commonly employed in mass spectrometry to analyze large biomolecules. It is particularly valuable for studying proteins, peptides and nucleic acids. During ESI, a high voltage is applied to a liquid sample containing the analyte, forming highly charged droplets. As the solvent evaporates, the analyte molecules are ionized and transferred into the gas phase. ESI typically generates multiply charged ions, allowing for precise molecular weight determination and enabling the sequencing and identification of biomolecules. ESI has revolutionized the field of proteomics and greatly facilitated the analysis of complex biological samples (Kliem, 2010).

- *Matrix-assisted laser desorption/ionization (MALDI)*

Matrix-Assisted Laser Desorption/Ionization (MALDI) is a powerful ionization technique widely employed in mass spectrometry to analyze large biomolecules. In MALDI, the analyte molecules are combined with a matrix compound, which assists in ionization. The analyte-matrix mixture is applied to a sample plate and irradiated with a laser beam. The laser energy causes the matrix to absorb the energy and rapidly vaporize, desorbing and ionizing the analyte molecules. The resulting ions are then detected and analyzed by the mass spectrometer. MALDI has significantly contributed to proteomics, enabling the analysis of high molecular weight compounds, such as proteins and peptides and facilitating biomolecule identification, characterization and structural elucidation (Figure 7.4) (Makarov, 2013).

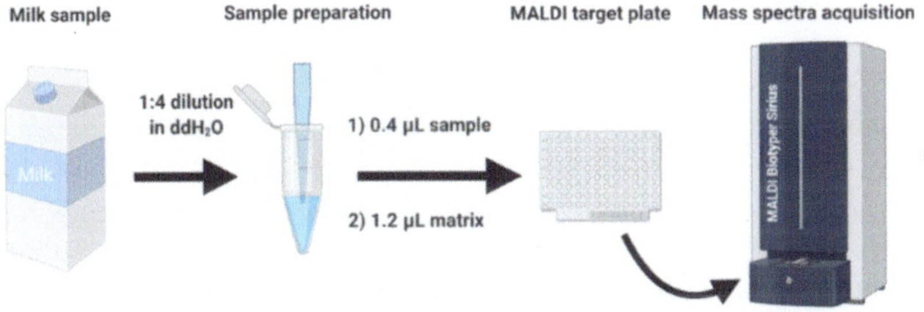

Figure 7.4. MALDI Sample Preparation.

Source: England, Philippa & Tang, Wenhao & Kostrzewa, Markus & Shahrezaei, Vahid & Gerald, Larrouy-Maumus. (2020). Discrimination of bovine milk from non-dairy milk by lipids fingerprinting using routine matrix-assisted laser desorption ionization mass spectrometry. Scientific Reports. 10. Scientific

Figure on ResearchGate. Available from: https://www.researchgate.net/figure/Schematic-diagram-of-the-sample-preparation-process-for-the-test-performed-on-the-MALDI_fig4_340060841 (accessed on 22 September 2023)

- *Atmospheric pressure chemical ionization (APCI)*

Atmospheric Pressure Chemical Ionization (APCI) is an ionization method commonly used in mass spectrometry to analyze polar and non-polar compounds. APCI operates at atmospheric pressure, making it suitable for direct analysis of samples without extensive sample preparation. In APCI, the analyte is introduced into the ion source, vaporized and ionized by a corona discharge or a corona needle (Figure 7.5). The ionization process typically results in protonated or deprotonated molecules, depending on the ionization conditions and the nature of the analyte. APCI is versatile and enables the analysis of a wide range of compounds, making it an essential tool in many applications, including pharmaceutical analysis, environmental monitoring and forensic investigations (Makarov, 2013).

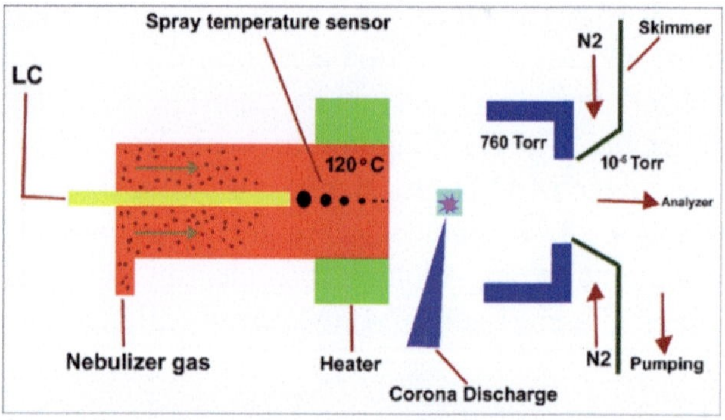

Figure 7.5. Atmospheric pressure chemical ionization source.

Source: Parmar, Jagdish & Chaubey, Kundan & Gupta, Vikas & Bharath, Manthena. (2021). Assessment of various veterinary drug residues in animal originated food products. Veterinary World. Scientific Figure on ResearchGate. Available from: https://www.researchgate.net/figure/Figure-6-Run-diagram-of-atmospheric-pressure-chemical-ionization-Source-Figure_fig5_352798218 (accessed on 22 September 2023)

Mass Analyzers

After ionization, various mass analyzers separate the ions based on their mass-to-charge ratio.

- *Magnetic sector analyzer*

A magnetic sector analyzer is a mass analyzer used in mass spectrometry. It utilizes a magnetic field to separate ions based on their momentum and charge. When ions enter the magnetic field, they follow curved paths due to the Lorentz force acting on them. Lighter ions experience a greater deflection than heavier ions, separating them. By adjusting the strength of the magnetic field, specific ions of interest can be directed towards the detector for detection and analysis. Magnetic sector analyzers offer high-resolution and precise mass measurements, making them valuable tools in various applications, such as isotope ratio determination, elemental analysis and accurate mass determination of compounds (Kliem, 2010).

- *Quadrupole mass analyzer*

A quadrupole mass analyzer is a widely used type of mass analyzer in mass spectrometry. It consists of four cylindrical rods arranged in a square pattern. A quadrupole field is created by applying appropriate radio frequency (RF) and direct current (DC) voltages to the rods. This field selectively allows ions of specific mass-to-charge ratios to pass through the analyzer while filtering out others.

The RF and DC voltages can be adjusted to scan through a range of mass-to-charge ratios, making quadrupole analyzers versatile for selective detection and monitoring experiments. Different ions can be transmitted or blocked by varying the voltages, allowing for precise control over the ions that reach the detector.

Quadrupole mass analyzers offer several advantages, including high sensitivity, selectivity and scanning capabilities. They are commonly used in various applications, such as pharmaceutical analysis, environmental monitoring and metabolomics. Quadrupole analyzers are also a key component in tandem mass spectrometry experiments, where they are used for precursor ion selection and collision-induced dissociation (Makarov, 2013).

- *Time-of-flight (TOF) mass analyzer*

It operates based on the principle that ions of different mass-to-charge ratios will travel at different speeds when subjected to a fixed electric field. TOF analyzers measure the time ions travel a fixed distance from the ion source to the detector.

Lighter ions, having a higher velocity, reach the detector faster, while heavier ions take longer to get to the sensor due to their lower rate. The mass-to-charge ratio of each ion can be determined by measuring the flight times of ions. TOF analyzers offer several advantages, including fast data acquisition rates, high mass range coverage and high resolution. They can rapidly analyze many ions and are well-suited for applications requiring high-speed analysis, such as proteomics, metabolomics and drug discovery. Additionally, TOF analyzers can handle complex mixtures and provide accurate mass measurements, making them valuable tools in various fields of research and analysis (Kliem, 2010).

- *Ion trap mass analyzer*

An ion trap mass analyzer uses an electromagnetic field to trap ions within a three-dimensional space. By varying the trapping voltages, ions of different mass-to-charge ratios can be selectively ejected and detected. Ion trap analyzers provide high sensitivity and the ability to perform tandem mass spectrometry experiments.

- *Fourier transform ion cyclotron resonance (FT-ICR) mass analyzer*

FT-ICR mass analyzers use strong magnetic and radio frequency electric fields to trap and detect ions. Ions are confined in a Penning trap and their motion induces an electric current that is detected. FT-ICR analyzers offer ultra-high resolution and mass accuracy and can efficiently analyze complex mixtures.

Ion detection and data acquisition

Once the ions are separated by the mass analyzer, their detection and subsequent data acquisition are critical for the analysis (Makarov, 2013).

- *Ion detectors*

Ion detectors convert ions into electrical signals that can be measured and analyzed. Several types of ion detectors are used, including:

Electron multipliers: Electron multipliers are widely used ion detectors in mass spectrometry that rely on the principle of secondary electron emission. When an ion strikes the surface of the detector, it imparts enough energy to release secondary electrons from the detector material. These secondary electrons are then accelerated by an electric field and directed towards a series of dynodes, metal surfaces with a high electron-emission coefficient. As the secondary electrons strike each dynode, more electrons are released, resulting in an amplification of the signal. This cascade of electron multiplication significantly increases the detected current, resulting in a signal proportional to the number of ions that struck the detector surface. Electron multipliers offer high sensitivity and are commonly used in applications requiring low detection limits and precise quantification in mass spectrometry (Kliem, 2010).

Faraday cups: Faraday cups are essential ion detectors used in mass spectrometry. They consist of metal cups that collect ions and measure the resulting ion current. When ions enter the Faraday cup, they deposit their charge on its surface, generating an electric current proportional to the number of ions collected. Faraday cups offer high sensitivity, enabling the detection of low levels of ions and they are commonly used for precise quantitative analysis in mass spectrometry. Their simplicity, reliability and linearity make Faraday cups an invaluable tool for various applications, including isotopic ratio measurements, elemental analysis and absolute quantification of analytes.

Channeltron detectors: *Channeltron* detectors are specialized electron multipliers commonly utilized as ion detectors in mass spectrometry. They consist of a microchannel plate (MCP) that acts as an electron multiplier. The MCP is a thin plate with microscopic channels that amplify the ion current. When ions strike the surface of the MCP, they release secondary electrons that are accelerated and multiplied as they pass through the channels.

This amplification process significantly increases the ion current, providing high sensitivity for ion detection. *Channeltron* detectors offer fast response times, allowing for rapid data acquisition and their high sensitivity makes them ideal for applications requiring low detection limits, such as trace analysis and high-resolution mass spectrometry (Kliem, 2010).

Microchannel plate detectors: They consist of a thin plate with an array of microscopic channels. When ions strike the surface of the MCP, they release secondary electrons that are accelerated and multiplied as they travel through the media. This multiplication process amplifies the ion signal, resulting in high sensitivity for ion detection (Makarov, 2013).

One of the key advantages of MCP detectors is their exceptional spatial resolution. The closely packed channels allow for precise localization of ion impacts, making MCP detectors suitable for imaging mass spectrometry applications. By scanning the sample surface, spatial distribution maps of ions can be generated, providing valuable information about the distribution of analytes in a sample.

Additionally, MCP detectors offer fast response times, allowing for rapid data acquisition, making them well-suited for time-critical experiments and high-throughput analysis. High sensitivity and spatial resolution make MCP detectors indispensable in surface analysis, materials science and biological imaging.

Data acquisition systems

Data acquisition systems process and record the ion signals detected by the ion detectors.

Analog-to-digital converters: Analogue-to-digital converters (ADCs) are vital components in mass spectrometry systems that transform analogue ion signals into digital signals. The ion signal, typically in the form of an electrical current or voltage, is continuously measured by the ADC at discrete time intervals. The ADC quantizes the analogue signal by sampling its intensity and converting it into a digital representation. This digital representation consists of a series of binary values that accurately represent the ion signal's amplitude at each sampling point. The resolution of the ADC determines the number of discrete levels that can be assigned to the analogue signal, affecting the fidelity and precision of the digitized data. The digital signal can then be further processed, stored and analyzed using computational techniques. ADCs play a critical role in capturing and preserving the ion signal information in a format that can be manipulated and interpreted by data acquisition systems and software algorithms in mass spectrometry (Makarov, 2013).

Time-to-digital converters: Time-to-digital converters (TDCs) are essential in time-of-flight (TOF) mass spectrometry systems. TDCs measure the time intervals between ion events, which is particularly relevant in TOF analyzers where the flight times of ions are measured. TDCs accurately measure the time ions travel a fixed distance from the ion source to the detector. By precisely capturing these flight times, TDCs determine the mass-to-charge ratio of ions based on their flight time.

TDCs convert the measured time intervals into digital values, allowing for precise and quantitative analysis of the ion flight times. These digital values can be further processed and analyzed to extract valuable information about the ions' masses and velocities. TDCs enable high-resolution and accurate mass measurements in time-of-flight mass spectrometry experiments (Kliem, 2010).

Data processing software: Data processing software analyzes and interprets the recorded ion signals. It performs tasks such as peak detection, spectral deconvolution and identification of compounds based on mass spectra. Data processing software also facilitates data visualization, mining and interpretation.

7.3. APPLICATIONS OF MASS SPECTROMETRY

Mass spectrometry has various applications in various fields, including analytical chemistry, forensic science, environmental monitoring and pharmaceutical research. This section explores the diverse applications of mass spectrometry and highlights some key areas where it has made significant contributions (Figure 7.6).

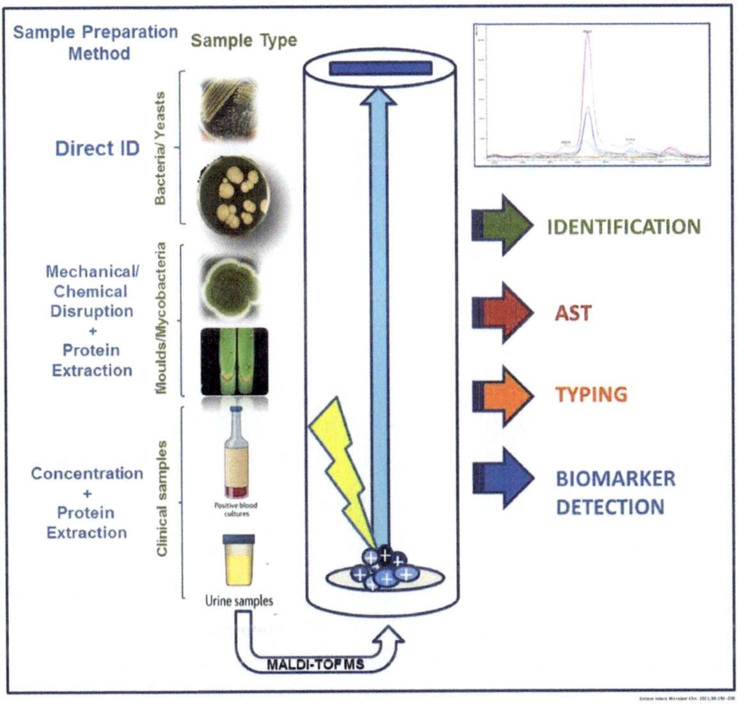

Figure 7.6. C-MS analysis of environmental samples.

Source: Oviaño, Marina & Rodriguez-Sanchez, Belen. (2020). MALDI-TOF mass spectrometry in the 21st century clinical microbiology laboratory. Enfermedades Infecciosas y Microbiología Clínica. 39. Scientific Figure on ResearchGate. Available from: https://www.researchgate.net/figure/Sample-types-that-can-be-analyzed-by-MALDI-TOF-and-the-processing-method-recommended_fig1_340942401 (accessed on 22 September 2023)

7.3.1. Analytical Applications

Mass spectrometry is a powerful analytical tool for identifying and quantifying compounds. It enables the analysis of complex mixtures, providing detailed information about the molecular composition and structure of the analytes. Several techniques have been developed to combine mass spectrometry with separation methods, such as gas chromatography (GC), liquid chromatography (LC), capillary electrophoresis (CE) and ambient ionization methods (Makarov, 2013).

Identification and quantification of compounds: The coupling of mass spectrometry with separation techniques has revolutionized chemical analysis by providing powerful tools for compound identification and quantification in complex matrices.Gas chromatography-mass spectrometry (GC-MS) is widely employed to analyze volatile and semi-volatile compounds. It offers high sensitivity and selectivity, allowing for detecting and quantifying analytes present at low concentrations. GC-MS separates compounds based on their volatility and affinity for the stationary phase, followed by their detection and identification using mass spectrometry (Kliem, 2010).

Liquid chromatography-mass spectrometry (LC-MS) extends the capabilities of mass spectrometry to a broader range of compounds, including polar and non-polar analytes. LC-MS utilizes liquid chromatography to separate compounds based on their interactions with the stationary phase, followed by their detection and characterization using mass spectrometry. LC-MS is highly versatile and finds applications in various fields, including pharmaceutical analysis, environmental monitoring and metabolomics.

Capillary electrophoresis-mass spectrometry (CE-MS) combines the separation power of capillary electrophoresis with the sensitivity and specificity of mass spectrometry. CE-MS separates compounds based on their charge-to-mass ratio and allows for detecting and identifying analytes in complex mixtures. It is particularly useful for analyzing charged compounds like peptides, proteins and nucleic acids.

Ambient mass spectrometry techniques, such as direct analysis in real-time (DART) and desorption electrospray ionization (DESI), enable rapid and straightforward analysis of samples in their native state without extensive sample preparation. These techniques offer advantages in terms of speed, simplicity and the ability to analyze samples in situ, making them valuable tools in applications such as food analysis, forensic investigations and environmental monitoring.

Combining mass spectrometry with separation techniques provides powerful analytical tools for identifying and quantifying compounds in various applications. The sensitivity, selectivity and versatility of these techniques have greatly advanced the field of chemical analysis, facilitating research in pharmaceuticals, environmental science and metabolomics.

Mass spectrometry plays a vital role in the structural analysis of biomolecules, including proteins, glycans, lipids and metabolites. Protein mass spectrometry allows protein identification, characterization and quantification, enabling proteomics and protein structure analysis studies. Glycan analysis utilizing mass spectrometry provides valuable insights into glycosylation patterns and their role in various biological processes. Lipidomic uses mass spectrometry to explore the composition and distribution of lipids in biological systems. Metabolomics aims to comprehensively analyze small molecules involved in cellular metabolism using mass spectrometry-based techniques, providing insights into metabolic pathways and disease biomarkers (Makarov, 2013).

7.3.2. Forensic Applications

Mass spectrometry has significant applications in forensic science, aiding in detecting, analyzing and characterizing substances of forensic interest.

Drug detection and analysis: It provides a highly specific and sensitive method for identifying and quantifying drugs and their metabolites in various biological samples, including urine, blood and hair.

One of the key advantages of mass spectrometry in drug analysis is its ability to provide high specificity. By utilizing tandem mass spectrometry (MS/MS), which involves multiple stages of mass analysis, achieving excellent selectivity in detecting specific drug compounds is possible. MS/MS can distinguish between drugs with similar molecular structures, reducing the risk of false positives or negatives (Kliem, 2010). Moreover, mass spectrometry offers exceptional sensitivity, allowing for the detection of drugs at very low concentrations. This is particularly important in drug testing scenarios where even trace amounts of drugs must be detected for accurate assessment. Forensic toxicology heavily relies on mass spectrometry to analyze samples from criminal investigations, postmortem examinations and workplace drug testing. The ability of mass spectrometry to detect a wide range of drugs, including illicit substances and prescription medications, contributes to the reliable identification and confirmation of drug presence. Furthermore, mass spectrometry can provide valuable information about drug metabolism and the presence of metabolites. Metabolite identification aids in understanding drug metabolism pathways, determining the extent of drug use and differentiating between recent and historical drug exposure.

Explosives detection: Mass spectrometry is extensively used in the detection and analysis of explosives. It enables the rapid and accurate identification of explosive compounds and their degradation products. Mass spectrometry techniques, such as ion mobility spectrometry (IMS) and ion trap mass analyzers, enhance the sensitivity and selectivity of explosive detection methods.

Forensic toxicology: Mass spectrometry plays a crucial role in forensic toxicology, enabling the detection and quantification of drugs, poisons and other toxic substances in biological samples. It provides vital information in cases of drug overdose, poisoning and postmortem investigations, aiding in determining the cause and manner of death (Makarov, 2013).

Environmental applications: Mass spectrometry widely monitors and identifies pollutants in air, water, soil and biological samples.

- ***Environmental Monitoring***

Environmental monitoring is a crucial application of mass spectrometry, enabling the detection, quantification and characterization of various pollutants in air, water, soil and biological samples. Mass spectrometry techniques play a vital role in assessing environmental contamination and understanding the sources and impacts of pollutants.

Mass spectrometry is widely employed in analyzing pesticides, commonly used in agricultural practices. It enables identifying and quantifying pesticide residues in food, water and soil, ensuring compliance with regulatory limits and assessing potential health risks (Figure 7.7) (Kliem, 2010).

Persistent organic pollutants (POPs) are a class of toxic compounds that are resistant to degradation and have long-term environmental impacts. Mass spectrometry techniques, such as high-resolution mass spectrometry (HRMS), are instrumental in identifying and quantifying POPs, such as polychlorinated biphenyls (PCBs) and dioxins, in environmental samples. This knowledge aids in understanding their distribution, transport and potential bioaccumulation in ecosystems. Polycyclic aromatic hydrocarbons (PAHs) are another group of environmental pollutants primarily generated from incomplete combustion processes. Mass spectrometry allows for the detection and quantification of PAHs in air, water and soil samples. These analyzes contribute to assessing the exposure risks and potential impacts on human health and the environment (Makarov, 2013).

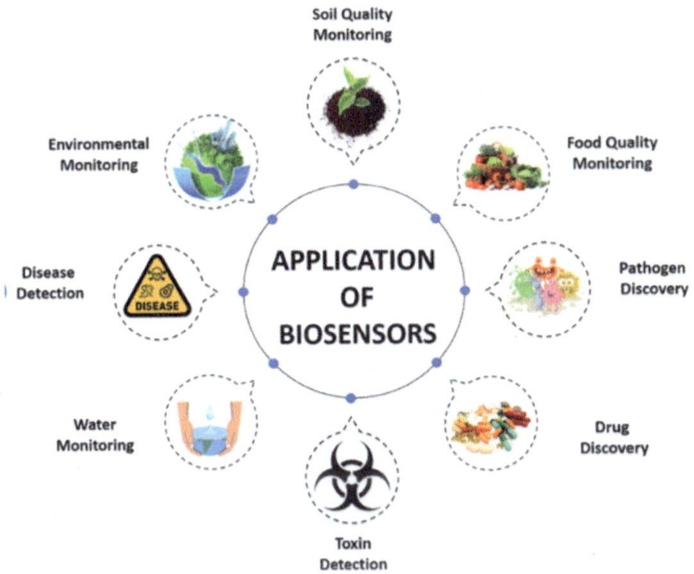

Figure 7.7. Applications of mass spectrometry in environmental monitoring using biosensors.

Source: Tetyana, Phumlani & Shumbula, Poslet & Njengele-Tetyana, Zikhona. (2021). Biosensors: Design, Development and Applications. Scientific Figure on ResearchGate. Available from: https://www.researchgate.net/figure/Various-applications-where-biosensors-have-been-used-Picture-adapted-from-76_fig3_352378597 (accessed on 22 September 2023)

Mass spectrometry techniques are also employed in analyzing heavy metals, such as lead, mercury and arsenic, which are widespread environmental pollutants with detrimental effects on ecosystems and human health. By accurately quantifying these heavy metals in environmental samples, mass spectrometry aids in evaluating contamination levels, assessing potential risks and implementing appropriate remediation measures (Kliem, 2010).

- **Identification of pollutants**

Mass spectrometry is crucial in identifying and characterizing emerging pollutants in environmental samples. These pollutants include pharmaceuticals, personal care products and endocrine-disrupting compounds. Mass spectrometry techniques offer the sensitivity and selectivity required to detect trace levels of these compounds, even in complex environmental matrices.

By accurately identifying and quantifying these emerging pollutants, mass spectrometry supports ecological risk assessments and informs regulatory measures to mitigate their potential impacts on ecosystems and human health. Mass spectrometry enables scientists to monitor the presence and levels of these pollutants, contributing to a better understanding of their distribution, fate and potential environmental risks.

7.3.3. Pharmaceutical Applications

Mass spectrometry is vital in pharmaceutical research and development, offering insights into drug development, characterization and pharmacokinetics.

Drug development and characterization: Mass spectrometry plays a vital role in drug development and characterization, offering valuable insights into various aspects of the drug discovery process. It provides essential information for understanding drug metabolism, identifying metabolites and characterizing drug candidates.

One important application of mass spectrometry in drug development is drug metabolism studies. Mass spectrometry identifies and characterizes metabolites formed during drug metabolism, providing insights into the biotransformation pathways and potential drug-drug interactions. This information is crucial for understanding a drug candidate's safety, efficacy and pharmacokinetics.

Metabolite identification is another key area where mass spectrometry excels. By analyzing biological samples, mass spectrometry can identify and characterize metabolites from drug metabolism. This aids in understanding the metabolic fate of a drug in the body and identifying any potentially toxic or active metabolites.

Mass spectrometry is also utilized to analyze drug impurities, unwanted substances that may be present in drug formulations. By accurately detecting and quantifying contaminants, mass spectrometry ensures the safety and purity of the drug product (Makarov, 2013).

Stability studies are essential to drug development to assess the shelf-life and degradation pathways of drugs. Mass spectrometry techniques, such as LC-MS, can monitor drug degradation products and impurities over time, providing valuable information on drug stability and formulation optimization. Additionally, mass spectrometry aids in characterizing drug candidates, including determining their molecular weight, structure and fragmentation patterns. This information is vital for verifying the identity and purity of drug candidates during the development process.

Pharmacokinetics and metabolism studies: One key application of mass spectrometry in pharmacokinetics is the measurement of drug concentrations in biological samples. By accurately quantifying drug levels over time, mass spectrometry allows researchers to determine key pharmacokinetic parameters such as clearance, volume of distribution and half-life. This information aids in assessing drug efficacy, optimizing dosing regimens and understanding the relationship between drug concentration and therapeutic response (Kliem, 2010).

Mass spectrometry also facilitates identifying and characterizing metabolic pathways involved in drug metabolism. Metabolites, which are products of drug metabolism, can be detected and identified using mass spectrometry. This information helps researchers understand the biotransformation of drugs and identify potential metabolites with pharmacological activity or toxicity.

Furthermore, mass spectrometry enables the identification of metabolites in biological samples. Mass spectrometry can identify and quantify metabolites generated from drug metabolism by analyzing samples such as urine or plasma. This knowledge contributes to understanding the metabolic fate of drugs in the body and identifying potential drug-drug interactions or adverse effects.

In addition to pharmacokinetics and metabolism studies, mass spectrometry aids in determining drug excretion pathways. Mass spectrometry can identify drug metabolites eliminated from the body by analyzing excreta samples, such as urine or faeces. This information assists in evaluating drug clearance mechanisms and optimizing drug formulations.

Mass spectrometry finds extensive applications in analytical chemistry, forensic science, environmental monitoring and pharmaceutical research. Its versatility, sensitivity and ability to provide structural information make it a valuable tool in various scientific disciplines, contributing to advancements in multiple fields and fostering a deeper understanding of complex biological, chemical and environmental systems (Makarov, 2013).

7.4. ADVANCES IN MASS SPECTROMETRY

Mass spectrometry has witnessed significant advancements in recent years, revolutionizing the field of analytical chemistry and expanding its applications across various disciplines. This section highlights some key advances in mass spectrometry and their impact on scientific research and analysis.

7.4.1. High-Resolution Mass Spectrometry

High-resolution mass spectrometry (HRMS) is a breakthrough advancement that has greatly enhanced mass spectrometry measurements' resolving power and accuracy. HRMS enables the determination of the mass-to-charge ratio of ions with exceptional precision, allowing for the identification and characterization of compounds at the molecular level (Kliem, 2010).

One of the key advantages of HRMS is its ability to differentiate between ions with similar masses but different elemental compositions. This capability is particularly important in complex samples, where the presence of isobaric compounds can complicate analysis. HRMS also facilitates the detection and quantification of trace-level analytes, providing enhanced sensitivity and selectivity. HRMS techniques, such as high-resolution time-of-flight (TOF) and orbitrap mass analyzers, have become instrumental in various fields, including proteomics, metabolomics, environmental analysis and pharmaceutical research (Figure 7.8). HRMS's high resolving power and mass accuracy enable the elucidation of complex molecular structures, identifying unknown compounds and discovering novel biomarkers.

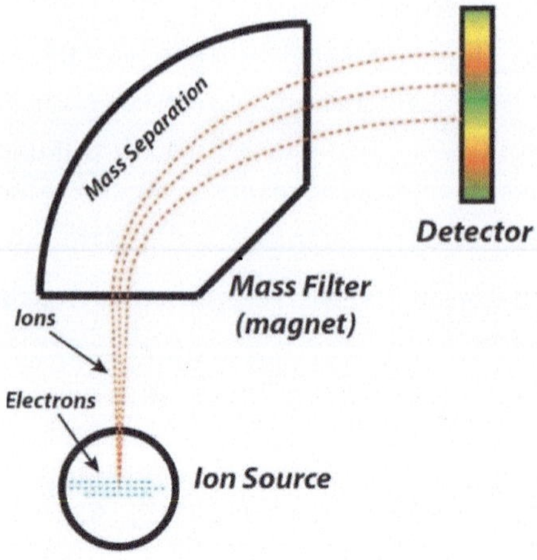

Figure 7.8. Schematic diagram of a mass spectrometer.

Source: Radauscher, Erich. (2015). Design, Fabrication and Characterization of Carbon Nanotube Field Emission Devices for Advanced Applications. Scien-

tific Figure on ResearchGate. Available from: https://www.researchgate.net/figure/Schematic-diagram-of-a-mass-spectrometer_fig14_312664731 (accessed on 22 September 2023)

7.4.2. Tandem Mass Spectrometry (MS/MS)

Tandem mass spectrometry, also known as MS/MS, combines multiple stages of mass analysis to provide detailed structural information about molecules. This advancement allows for the fragmentation of ions and the study of their fragment patterns, aiding in compound identification and structural elucidation (Makarov, 2013).

MS/MS techniques, such as collision-induced dissociation (CID) and electron capture dissociation (ECD), enable the generation of fragment ions by subjecting precursor ions to specific energy conditions. The resulting fragment ions can provide valuable insights into the analyte's molecular structure, functional groups and connectivity.

The combination of MS/MS with high-resolution mass spectrometry further enhances the capabilities of both techniques. High-resolution MS/MS allows for the precise determination of fragment ion masses, improving the accuracy of structural elucidation and compound identification (Kliem, 2010).

MS/MS is widely used in proteomics, metabolomics and small molecule analysis, enabling the characterization of complex mixtures and the identification of biomolecules with high confidence. It has become an indispensable tool in peptide sequencing, post-translational modification analysis and the determination of metabolic pathways.

7.4.3. Imaging Mass Spectrometry

Imaging mass spectrometry (IMS) is an innovative technique that enables the visualization and spatial mapping of molecules within a sample. It combines the power of mass spectrometry with imaging capabilities, allowing researchers to study the distribution and localization of specific compounds within biological tissues, cells and surfaces.

IMS techniques, such as matrix-assisted laser desorption/ionization (MALDI) imaging, desorption electrospray ionization (DESI) imaging and secondary ion mass spectrometry (SIMS), offer valuable insights into the spatial organization of molecules in complex samples. By generating molecular maps, IMS provides information about the distribution of metabolites, lipids, peptides and other biomolecules within biological systems.

IMS has applications in various fields, including biomedical research, pharmacology, environmental science and materials analysis. It has contributed to the identification of disease biomarkers, the evaluation of drug distribution and penetration, the assessment of chemical composition in geological samples and the analysis of surface properties.

Integrating IMS with other analytical techniques, such as high-resolution mass spectrometry and molecular imaging modalities (e.g., optical microscopy or magnetic resonance imaging), further enhances its capabilities. This multimodal imaging approach enables the correlation of molecular information with cellular structures, providing comprehensive insights into biological processes and disease mechanisms (Makarov, 2013).

7.4.4. Proteomics and Peptide Sequencing

Mass spectrometry has revolutionized proteomics, the large-scale study of proteins, by enabling accurate protein identification, quantification and characterization. Analyzing complex protein mixtures and determining their primary structures has significantly advanced our understanding of biological systems.

Proteomics relies on mass spectrometry techniques, such as liquid chromatography-mass spectrometry (LC-MS) and tandem mass spectrometry (MS/MS), for protein identification and peptide sequencing. LC-MS allows for the separation of complex protein mixtures, while MS/MS enables the fragmentation and analysis of peptides, leading to the determination of their amino acid sequences (Kliem, 2010).

Advanced mass spectrometry-based proteomics approaches, such as shotgun proteomics and targeted proteomics, have emerged as powerful tools in biomarker discovery, protein quantification and post-translational modification analysis. These techniques enable comprehensive profiling of protein expression, dynamic changes in protein abundance and the identification of modified peptides.

Furthermore, advancements in mass spectrometry have expanded proteomics to include protein interaction studies, structural analysis and protein subcellular localization. Cross-linking mass spectrometry (XL-MS) and hydrogen/deuterium exchange mass spectrometry (HDX-MS) provide insights into protein-protein interactions, folding and dynamics.

7.4.5. Metabolomics and Lipidomics

Metabolomics and *lipidomics* are rapidly growing fields that aim to analyze metabolites and lipids in biological systems comprehensively. Mass spectrometry plays a pivotal role in these disciplines, enabling the identification, quantification and profiling of small molecules in various samples, including biofluids, tissues and cell extracts.

Metabolomics utilizes mass spectrometry techniques, such as gas chromatography-mass spectrometry (GC-MS) and liquid chromatography-mass spectrometry (LC-MS), to analyze metabolites involved in metabolic pathways, signaling and cellular processes. Mass spectrometry-based metabolomics allows for the characterization of metabolic profiles, the identification of biomarkers and the investigation of metabolic dysregulation in diseases (Makarov, 2013).

Lipidomics focuses on the comprehensive analysis of lipid species, including fatty acids, *glycerolipids,* phospholipids and sphingolipids. Mass spectrometry, particularly shotgun *lipidomics* and lipid profiling approaches, enables the identification and quantification of lipid species with high precision and sensitivity. This information contributes to understanding lipid metabolism, lipid signaling pathways and lipid-related diseases.

Advanced mass spectrometry techniques, such as high-resolution mass spectrometry (HRMS), have significantly expanded the coverage and accuracy of metabolomics and *lipidomics* analyzes. HRMS facilitates the identification and differentiation of isomeric metabolites and lipid species, allowing for more comprehensive and precise metabolite profiling.

Moreover, mass spectrometry-based imaging techniques, such as MALDI and DESI imaging, enable the spatial visualization of metabolites and lipids within biological samples. This capability provides insights into their distribution patterns, cellular localization and metabolic dynamics (Kliem, 2010).

7.4.6. Ambient Ionization Techniques

Ambient ionization techniques have emerged as powerful tools in mass spectrometry, enabling direct analysis of samples in their native state without extensive sample preparation. These techniques offer rapid analysis, high sensitivity and the ability to analyze various models, including solids, liquids and gases. Techniques such as direct analysis in real time (DART), desorption electrospray ionization (DESI) and laser ablation electrospray ionization (LAESI) are examples of ambient ionization techniques. They use various mechanisms to generate ions directly from the sample surface or the surrounding environment, facilitating rapid analysis and real-time monitoring. Ambient ionization techniques find applications in areas such as forensics, environmental analysis, food safety and clinical diagnostics. They enable on-site analysis, rapid screening of samples and non-destructive analysis of precious illustrations. These techniques have expanded the accessibility and versatility of mass spectrometry, allowing for in situ analysis in diverse settings.

7.4.7. Ion Mobility Spectrometry Coupled with Mass Spectrometry

Ion mobility spectrometry (IMS) is a separation technique that measures the time it takes for ions to traverse a gas-filled drift tube under the influence of an electric field. When coupled with mass spectrometry, IMS enhances the separation power and selectivity of mass spectrometry measurements.

IMS provides an additional dimension of separation based on the size, shape and charge of ions. It complements the mass-to-charge ratio separation in mass spectrometry, enabling the separation of isomeric compounds and the identification of structural isomers.

By measuring the ion mobility, which reflects the ion's ability to move through the gas phase, IMS offers valuable information about molecules' three-dimensional structure and conformation (Makarov, 2013). The combination of IMS with mass spectrometry, known as IMS-MS, has metabolomics, proteomics and drug discovery applications. IMS-MS enhances the characterization of complex mixtures, improves the identification of unknown compounds and provides additional insights into compound structures and interactions (Kliem, 2010).

IMS-MS has also been applied in analyzing complex biological samples, such as protein complexes and protein-ligand interactions. It aids in the elucidation of protein conformations, the characterization of protein complexes and the investigation of protein folding and dynamics.

7.5. CHALLENGES AND FUTURE DIRECTIONS IN MASS SPECTROMETRY

Mass spectrometry has made significant strides in recent years, but there are still challenges to overcome and exciting possibilities on the horizon.

This section highlights some of the key challenges in mass spectrometry and explores potential future directions for the field.

7.5.1. Instrumentation Challenges

One of the ongoing challenges in mass spectrometry is the development of advanced instrumentation to meet the increasing demands for sensitivity, resolution and speed. Instrumentation improvements are crucial for expanding the capabilities of mass spectrometry in various applications, such as proteomics, metabolomics and environmental analysis.

Enhancing the sensitivity of mass spectrometry is particularly important for analysing trace-level compounds, biomarkers and low-abundance species. Improvements in ionization techniques, such as developing novel ion sources and improved interface designs, can help increase ionization efficiency and sensitivity (Kliem, 2010).

Another challenge is improving the resolution and accuracy of mass spectrometry measurements. High-resolution mass spectrometry (HRMS) has already made significant advancements in this area, but further developments in mass analyzers, ion optics and detection systems are needed to push the boundaries of resolution and mass accuracy.

Furthermore, there is a need for more compact and cost-effective mass spectrometry instruments that can be widely accessible in various settings, such as point-of-care diagnostics, field analysis and resource-limited environments. Miniaturization and portability of mass spectrometry systems are active research areas, enabling on-site analysis, rapid screening and real-time monitoring (Makarov, 2013).

7.5.2. Data Analysis and Interpretation

With the increasing complexity and volume of data generated by mass spectrometry, there is a pressing need for improved data analysis and interpretation strategies. The interpretation of mass spectrometry data requires sophisticated algorithms, statistical tools and bioinformatics approaches to extract meaningful information and translate raw data into biological insights.

Data analysis challenges include peak picking, deconvolution, alignment, normalization and identifying unknown compounds. Integrating mass spectrometry data with other omics data, such as genomics and transcriptomics, adds another layer of complexity and necessitates the development of integrated data analysis strategies. Machine learning and artificial intelligence (AI) approaches are promising in addressing these data analysis challenges. These techniques can aid in pattern recognition, feature selection, classification and prediction, enhancing data analysis and interpretation efficiency and accuracy in mass spectrometry.

7.5.3. Biomolecular Complexities

Mass spectrometry faces challenges in dealing with the inherent complexities of biomolecular systems, such as proteins, peptides, metabolites and lipids. These complexities include post-translational modifications, structural heterogeneity, dynamic modifications and isomeric compounds.

Addressing these challenges requires advancements in mass spectrometry techniques that enable the comprehensive characterization of biomolecules. This includes improved fragmentation methods, such as electron capture dissociation (ECD) and electron transfer dissociation (ETD), which can facilitate the analysis of post-translational modifications and structural elucidation of peptides and proteins (Kliem, 2010).

The development of ion mobility spectrometry (IMS) coupled with mass spectrometry offers the potential to address the challenges of biomolecular complexity. IMS can provide additional information about molecular size, shape and conformation, enabling better differentiation and characterization of isomeric compounds and complex biomolecular assemblies.

7.5.4. Miniaturization and Portable Mass Spectrometry

Miniaturization and portable mass spectrometry systems are promising directions for the future of the field. The ability to perform mass spectrometry analysis in compact, mobile devices opens up new opportunities for on-site research, field analysis and point-of-care diagnostics (Makarov, 2013).

Miniaturization efforts aim to reduce mass spectrometers' size, weight and power requirements without compromising performance. This involves the development of miniaturized ion sources, mass analyzers and detectors, as well as advances in microfluidics and chip-based technologies.

Portable mass spectrometry systems offer advantages such as rapid analysis, real-time monitoring and the ability to analyze samples in their native state without extensive sample preparation. These systems can find applications in environmental monitoring, food safety, forensic analysis, clinical diagnostics and space exploration.

7.5.5. Integration with Other Analytical Techniques

Integrating mass spectrometry with other analytical techniques is a promising direction for future advancements. Combining mass spectrometry with complementary approaches allows a more comprehensive and multi-dimensional analysis of complex samples. Integration with chromatographic techniques, such as liquid chromatography (LC) and gas chromatography (GC), enhances mass spectrometry's separation power and analyte specificity. This enables the analysis of complex mixtures and provides more robust identification and quantification of analytes (Kliem, 2010).

Integrating mass spectrometry with imaging techniques, such as microscopy and spectroscopy, offers the potential for spatially resolved molecular analysis. Imaging mass spectrometry, for example, allows for the visualization and mapping of molecular distributions within biological samples, contributing to a deeper understanding of tissue heterogeneity and disease processes.

Moreover, combining mass spectrometry with emerging technologies, such as microfluidics, nanotechnology and biosensors, holds promise for new applications and improved analytical performance. These integrations can enhance sample handling, increase sensitivity and enable high-throughput analysis in various fields (Makarov, 2013).

7.5.6. Emerging Applications and Technologies

Mass spectrometry continues to expand its applications into new areas and benefit from emerging technologies. The field is witnessing the emergence of novel applications and methodologies driven by advances in sample preparation, ionization techniques and mass analyzers.

Emerging applications include single-cell analysis, microbiome analysis, spatial metabolomics and real-time monitoring of dynamic processes. These applications require developing specialized sample preparation methods, improved sensitivity and resolution and the ability to handle small sample volumes (Figure 7.9) (Kliem, 2010).

Technological advancements, such as ambient ionization techniques, high-throughput mass spectrometry and integrating mass spectrometry with imaging modalities, are pushing the boundaries of analytical capabilities and enabling discoveries (Makarov, 2013).

Additionally, integrating mass spectrometry with other omics technologies, such as genomics, transcriptomics and proteomics, offers a more comprehensive understanding of biological systems and enables systems-level analysis.

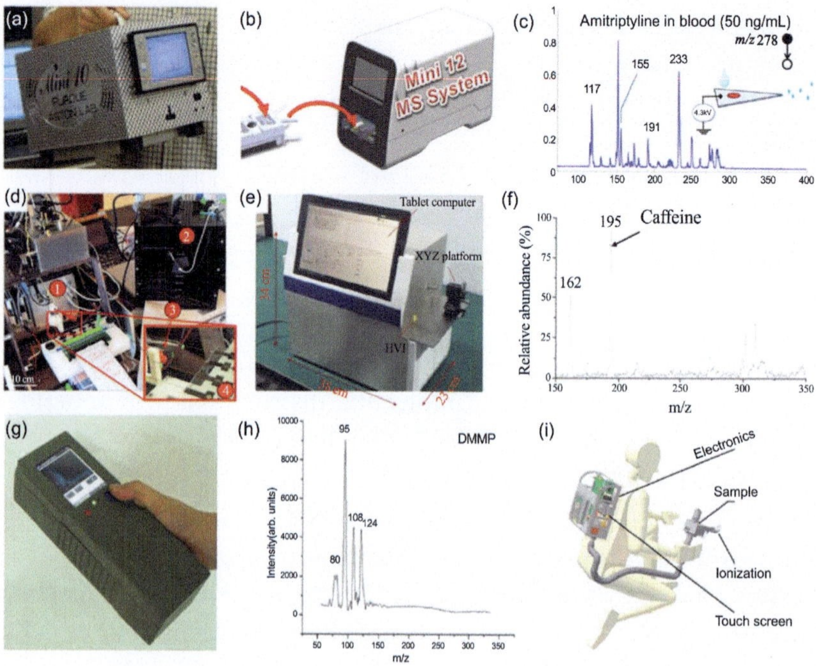

Figure 7.9. Miniaturized mass spectrometry device.

Source: Coskun, Ahmet & Topkaya, Seda & Yetisen, Ali & Cetin, Arif. (2018). Portable Multiplex Optical Assays. Advanced Optical Materials. Scientific Figure on ResearchGate. Available from: https://www.researchgate.net/figure/Mass-spectrometers-a-Handheld-mass-spectrometer-Mini-10-with-rectilinear-ion-trap_fig6_329397375 (accessed on 22 September 2023)

In conclusion, mass spectrometry is a powerful analytical technique that has revolutionized the field of scientific research and analysis. It offers various applications across various disciplines, including chemistry, biology, medicine, environmental science and forensics.

This chapter has provided a comprehensive overview of mass spectrometry, covering its principles, techniques, instrumentation and applications. We explored the different ionization methods, mass analyzers, ion detectors and data acquisition systems used in mass spectrometry. Additionally, we delved into the analytical applications of mass spectrometry, such as compound identification and quantification, structural analysis of biomolecules, forensic analysis, environmental monitoring and pharmaceutical research.

Furthermore, we discussed the advances in mass spectrometry, including high-resolution mass spectrometry, tandem mass spectrometry (MS/MS), imaging mass spectrometry and the emerging fields of proteomics, metabolomics and lipidomics. These advancements have expanded the capabilities of mass spectrometry, allowing for more accurate analysis, improved structural elucidation and comprehensive profiling of complex samples (Makarov, 2013).

However, mass spectrometry still faces challenges and has room for further development. Instrumentation challenges, such as enhancing sensitivity, resolution and miniaturization, are areas of ongoing research. Data analysis and interpretation strategies need to keep pace with the increasing complexity and volume of data generated by mass spectrometry. Biomolecular complexities, such as post-translational modifications and structural heterogeneity, pose challenges that require innovative approaches and techniques.

The future of mass spectrometry holds promising directions, including miniaturized and portable mass spectrometry systems, integration with other analytical techniques, advancements in data analysis and interpretation and exploration of emerging applications and technologies. These advancements will enable more comprehensive analysis, deeper insights into biological systems and enhanced capabilities for solving complex scientific challenges (Makarov, 2013).

CHAPTER 8

Nanopore-Based Measurements

Contents

8.1. INTRODUCTION

Nanopore technology has emerged as a powerful and versatile tool for a wide range of scientific and technological applications. From biophysics to genomics and beyond. By utilizing the unique properties of nanopores, scientists have unlocked new possibilities for molecular analysis and detection, opening up avenues for innovative research and applications (Radenovic, 2019) The concept of nanopores revolves around the use of nanoscale pores, typically just a few nanometers in diameter, as miniature sensors capable of analyzing individual molecules. These pores can be formed naturally, such as the ion channels found in biological membranes, or artificially engineered using advanced nanofabrication techniques. The key feature of nanopores lies in their ability to selectively allow molecules to pass through based on their size, shape, or charge, making them ideal tools for precise molecular characterization (Edel, 2020). The historical development of nanopore technology can be traced back to the mid-1990s when the idea of using nanopores for molecular sensing was first proposed. It was Dr. David Deamer and his colleagues who initially envisioned the potential of nanopores as a means to analyze DNA molecules.

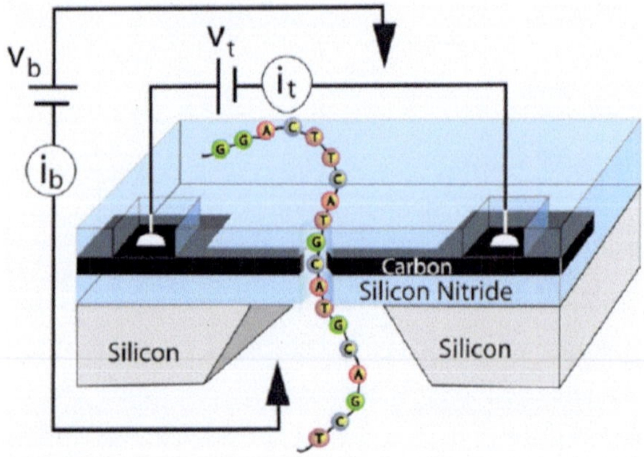

Figure 8.1. Diagram of the nanopore and measurement circuitry showing both blocking current, i b and transverse current.

Source: Spinney, Patrick & Collins, Scott & Howitt, David & Smith, Rosemary. (2012). Fabrication and Characterization of a Solid State Nanopore with Self-Aligned Carbon Nanoelectrodes for Molecular Detection. Nanotechnology. Scientific Figure on ResearchGate. Available from: https://www.researchgate.net/figure/Diagram-of-the-nanopore-and-measurement-circuitry-showing-both-blocking-current-i-b_fig1_221965104 (accessed 22 September 2023)

Their vision was based on the fundamental principle that as DNA strands pass through a nanopore, they would cause detectable changes in the electrical current flowing through the pore. This concept laid the foundation for nanopore-based DNA sequencing, which has since become one of the most prominent applications of the technology (Figure 8.1).

Over the years, significant advancements have been made in nanopore technology, driven by the convergence of interdisciplinary research efforts. Researchers from diverse fields, including physics, chemistry, materials science and biology, have contributed to refining nanopore fabrication techniques, improving measurement protocols and enhancing data analysis methods. As a result, nanopore technology has evolved from its humble beginnings as a proof-of-concept to a mature and versatile tool with broad applications (Radenovic, 2019).

One of the major breakthroughs in the field was the development of solid-state nanopores, which eliminated the reliance on fragile biological membranes. Solid-state nanopores are fabricated by etching nanoscale holes in thin, solid materials, such as silicon nitride or graphene. This approach offers several advantages, including greater stability, reproducibility and the ability to integrate nanopore sensors into microfluidic devices, enabling high-throughput measurements (Edel, 2020).

8.2. TYPES OF NANOPORES

In the field of nanopore-based measurements, a variety of nanopores have been developed and utilized to achieve diverse applications in nanoscience, biophysics and bioengineering. Nanopores are tiny openings, typically on the nanometer scale, that allow the passage of molecules or ions through them. They act as miniature channels, enabling the analysis and manipulation of individual molecules at the nanoscale. Understanding the different types of nanopores is crucial for harnessing their unique properties and tailoring them to specific experimental needs (Radenovic, 2019).

8.2.1. Biological Nanopores

Biological nanopores are protein-based channels found in living organisms, ranging from bacteria to higher organisms like humans. These nanopores are typically embedded within the lipid bilayer membranes of cells and play critical roles in biological processes such as molecular transport, signaling and ion homeostasis. Researchers have recognized the potential of biological nanopores for their ability to selectively translocate molecules, providing a basis for developing nanopore-based sensing and analysis platforms (Figure 8.2).

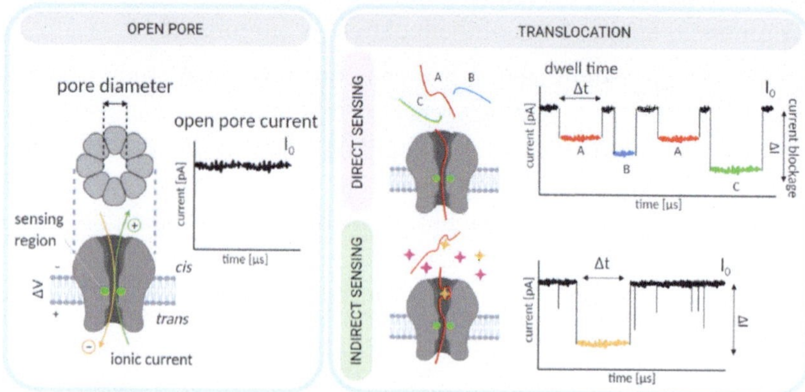

Figure 8.2. An illustration of a biological nanopore and the principle of nanopore-based biosensing.

Source: Crnković, A., Srnko, M., & Anderluh, G. (2021). Biological Nanopores: Engineering on Demand. Life, 11(1), 27. MDPI AG. Retrieved from http://dx.doi.org/10.3390/life11010027

One of the most well-known biological nanopores is the α-hemolysin pore, derived from the bacterium *Staphylococcus aureus*. This nanopore consists of a barrel-shaped structure formed by a heptameric assembly of α-hemolysin proteins. The α-hemolysin pore exhibits a characteristic constriction region, known as the vestibule, which allows the passage of single-stranded DNA or RNA molecules. By applying an electrical potential across the nanopore and monitoring the resulting ionic current, researchers can obtain valuable information about the translocating molecule, such as its size, shape and even sequence (Radenovic, 2019).

Another significant class of biological nanopores is the family of protein channels known as ion channels. Ion channels play a crucial role in regulating the flow of ions across cell membranes, maintaining the electrical potential and facilitating numerous physiological processes. These channels have been extensively studied and have shown great potential as nanopore sensors.

For instance, the bacterial protein channel, called Mycobacterium smegmatis porin A (MspA), has garnered significant interest in the field of nanopore research. MspA is an octameric channel that exhibits a funnel-shaped structure with a narrow constriction region. Researchers have found that MspA can be used to discriminate between various nucleotides, providing a platform for DNA sequencing and analysis (Edel, 2020).

Biological nanopores offer several advantages over other types of nanopores. First, their natural occurrence means that they can be readily obtained from biological sources, simplifying the fabrication process. Additionally, biological nanopores often exhibit high specificity and selectivity for certain molecules, making them ideal for molecular sensing and characterization. Moreover, these nanopores can be genetically modified or engineered to enhance their functionality and tailor them for specific applications.

In recent years, advancements in protein engineering and synthetic biology techniques have enabled the creation of novel biological nanopores with enhanced properties and functionalities. Scientists have successfully engineered nanopores to exhibit improved sensitivity, enhanced translocation speed and even the ability to detect specific analytes with high accuracy. These developments hold tremendous potential for the future of nanopore-based measurements and their integration into a wide range of applications, including DNA sequencing, protein analysis, drug screening and biomolecular detection (Edel, 2020).

8.2.2. Solid-State Nanopores

Another prominent category of nanopores is solid-state nanopores. These nanopores are fabricated using solid materials, typically synthetic materials with well-defined geometries, to create a controlled and predictable channel for molecule detection and analysis. Solid-state nanopores offer several advantages over other types of nanopores, such as biological or hybrid nanopores. Their robustness, stability and controllable dimensions make them highly suitable for a wide range of applications (Radenovic, 2019).

- ***Synthetic Nanopores***

Within the realm of solid-state nanopores, synthetic nanopores play a crucial role. Synthetic nanopores are artificially created structures designed with precise dimensions and characteristics to facilitate specific types of measurements and analyzes. These nanopores are typically fabricated using advanced nanofabrication techniques, such as electron beam lithography or focused ion beam milling, which allow for precise control over the shape and size of the nanopore (Figure 8.3).

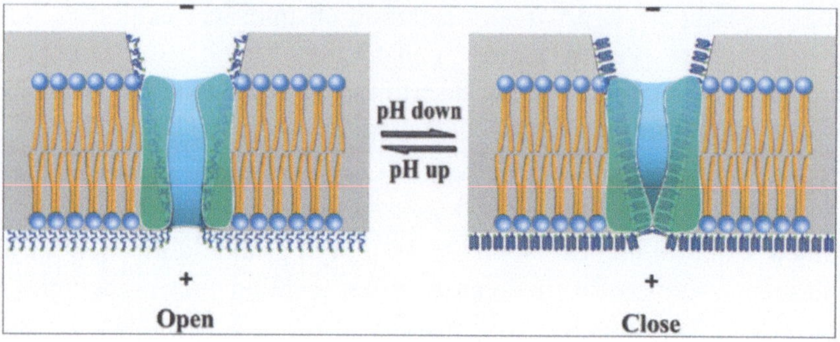

Figure 8.3. Single synthetic nanopores by proton-driven DNA molecular motors.

Source: Fan Xia, Wei Guo, Youdong Mao, Xu Hou, Jianming Xue, Hongwei Xia, Lin Wang, Yanling Song, Hang Ji, Qi Ouyang, Yugang Wang and Lei Jiang, (2008), Journal of the American Chemical Society, 130 (26), 8345–8350 https://doi.org/10.1021/ja800266p

Synthetic nanopores offer several advantages over natural or biological nanopores. Firstly, their fabrication process allows for customization of the nanopore's physical properties, such as diameter, length and surface chemistry, to match the specific requirements of the desired application. This control over the nanopore's characteristics enables fine-tuning of the detection sensitivity and selectivity for different types of molecules (Edel, 2020).

Secondly, synthetic nanopores can be engineered to be highly stable and durable, ensuring long-term functionality and reusability. Compared to biological nanopores, which may degrade or become unstable over time, synthetic nanopores provide a consistent and reliable platform for measurements.

Furthermore, synthetic nanopores can be integrated into various devices and systems, such as microfluidic platforms or biosensors, allowing for seamless integration with other components and enabling complex experimental setups. Their compatibility with complementary technologies and techniques enhances the versatility and applicability of synthetic nanopores in different fields (Radenovic, 2019).

Synthetic nanopores find applications in a wide range of scientific and technological areas. They have been extensively utilized in DNA sequencing, where the translocation of DNA molecules through the nanopore is measured to decode the genetic information. Synthetic nanopores also play a crucial role in single-molecule analysis, nanoparticle characterization and drug delivery systems, among other applications.

- ***Carbon nanotubes as nanopores***

Carbon nanotubes (CNTs) are cylindrical structures composed of carbon atoms arranged in a hexagonal lattice. They can be single-walled or multi-walled, depending on the number of concentric graphene layers forming the tube. The diameter of a carbon nanotube can vary from a few angstroms to several tens of nanometers, while their length can range from micrometers to millimeters.

When appropriately modified, carbon nanotubes can act as highly sensitive and selective nanopores. Their sub-nanometer dimensions make them suitable for single-molecule analysis, allowing the detection and characterization of various biomolecules such as DNA, RNA, proteins and small molecules. Carbon nanotubes offer several advantages as nanopores, including their robustness, chemical stability and electrical conductivity.

One of the key advantages of using carbon nanotubes as nanopores is their ability to discriminate between different molecules based on their size and charge. The diameter of a carbon nanotube can be tailored to match the size of the target molecule, enabling the precise control of molecular translocation through the nanopore. Additionally, the charged nature of the nanotube's surface can influence the movement of charged molecules, allowing for the detection of specific analytes with high sensitivity (Edel, 2020).

The fabrication of carbon nanotube nanopores involves several techniques, including direct insertion, dielectric breakdown and track-etched methods. Direct insertion involves inserting a nanotube into a preformed pore, while dielectric breakdown creates nanopores by applying a high electric field to a thin insulating membrane with embedded carbon nanotubes. Track-etched methods rely on etching a polymer membrane with ion bombardment to create nanopores, which are then coated with carbon nanotubes (Radenovic, 2019).

Carbon nanotube nanopores offer several potential applications across different fields. In DNA sequencing, for instance, they can be utilized to pass DNA strands through the nanopore and by measuring the changes in ionic current as the DNA translocates, the DNA sequence can be determined. Moreover, carbon nanotube nanopores can be used in biosensing applications, enabling the detection and analysis of biomarkers for disease diagnostics, environmental monitoring and drug discovery. Despite their significant potential, there are still challenges to overcome in utilizing carbon nanotubes as nanopores. Achieving controlled and uniform diameters, as well as precise positioning of nanotubes within the nanopores, remains a technical hurdle. Additionally, improving the signal-to-noise ratio and enhancing the reliability and reproducibility of measurements are areas of active research.

8.3. NANOPORE FABRICATION TECHNIQUES

The fabrication of nanopores involves precise engineering processes to create nanoscale structures with desired dimensions, shapes and properties. Different fabrication techniques have been developed over the years, each offering unique advantages and challenges. Let's delve into some of the prominent techniques used in nanopore fabrication:

8.3.1. Biological Nanopore Fabrication

One of the most intriguing and widely explored methods is biological nanopore fabrication. In this technique, nature's own nanopores, typically derived from proteins or DNA, are harnessed and modified for experimental purposes (Edel, 2020).

The biological nanopore fabrication process begins by identifying naturally occurring nanopores that exhibit desirable characteristics for a given application. These nanopores are often found in biological systems, such as ion channels in cell membranes or viral capsids. By carefully studying and understanding the properties of these natural nanopores, scientists can harness their unique abilities and integrate them into artificial systems (Radenovic, 2019).

Once a suitable biological nanopore is selected, it is isolated and extracted from its native source. This extraction process involves meticulous purification techniques to ensure the nanopore is free from impurities that could interfere with subsequent experiments. By isolating the nanopore, researchers gain control over its environment and can manipulate its structure to enhance its functionality.

To modify the biological nanopore for specific measurements, researchers employ various techniques, such as genetic engineering or chemical modifications. Genetic engineering allows for the production of recombinant proteins with tailored nanopore properties. By manipulating the genetic code, researchers can introduce desired mutations or amino acid substitutions to fine-tune the nanopore's characteristics, such as size, charge, or selectivity.

Chemical modifications provide an alternative method for nanopore customization. By treating the nanopore with specific molecules or altering its environment, researchers can introduce changes that enhance its performance. These modifications can include the attachment of functional groups, alteration of surface charge, or the addition of artificial gating mechanisms (Edel, 2020).

The next step in biological nanopore fabrication involves integrating the modified nanopore into an experimental setup. This can be achieved by embedding the nanopore within a lipid bilayer or solid-state membrane, allowing for controlled measurements and interactions with target molecules. The lipid bilayer technique involves forming a thin membrane composed of lipids, creating a stable environment for the nanopore to function. Alternatively, solid-state nanopores can be created using materials such as silicon nitride, graphene, or other nanoscale membranes, providing a robust platform for nanopore measurements.

Once integrated, the biological nanopore is ready for measurements. These measurements can include the translocation of molecules through the nanopore, electrical sensing of molecules, or spectroscopic analysis. By carefully monitoring the changes in current, voltage, or other physical properties as molecules interact with the nanopore, researchers can gather valuable information about the molecules' characteristics, such as size, charge, or conformation (Radenovic, 2019).

8.3.2. Solid-State Nanopore Fabrication

Solid-state nanopore fabrication stands out as one of the most widely adopted techniques for creating nanopores with exceptional precision and control. This method involves the creation of nanopores in solid-state materials, such as silicon nitride or graphene, through a series of intricate processes.

To begin the fabrication process, a suitable substrate material, often silicon, is carefully chosen based on its desirable properties, such as mechanical strength, thermal stability and compatibility with subsequent processing steps. The substrate is then cleaned meticulously to ensure the removal of any contaminants that could affect the nanopore formation (Edel, 2020).

Next, lithography techniques, such as electron beam lithography or photolithography, are employed to define the desired shape and size of the nanopore on the substrate. These lithographic methods enable the creation of precise nanopore patterns by selectively exposing the substrate to focused beams of electrons or ultraviolet light through a mask. The exposed areas undergo chemical or physical changes, forming a pattern that serves as a template for the nanopore fabrication.

Once the nanopore pattern is defined, various etching techniques come into play to remove the material surrounding the desired pore region. Reactive Ion Etching (RIE) or Focused Ion Beam (FIB) milling are commonly utilized methods, capable of selectively removing substrate material to create well-defined nanopores. These etching processes offer excellent control over the nanopore diameter and shape, crucial factors that influence the subsequent sensing capabilities of the nanopore device.

After the initial nanopore structure is formed, additional steps may be employed to further enhance the nanopore properties. For instance, the pore walls can be functionalized with specific molecules or coatings to modify the surface properties, making them more suitable for specific sensing applications. This functionalization step allows researchers to tailor the nanopore's interactions with target molecules, enhancing the sensitivity and selectivity of the device (Radenovic, 2019).

Finally, the fabricated solid-state nanopore is integrated into a sensing platform, such as a microfluidic device, to facilitate the controlled movement of molecules through the pore. This integration enables precise control over the experimental conditions, such as buffer composition, applied voltage and temperature, ensuring optimal measurement conditions for reliable and accurate data acquisition.

Solid-state nanopore fabrication is a highly sophisticated technique that offers precise control over the creation of nanopores in solid-state materials. Through a combination of lithography, etching and functionalization steps, researchers can produce well-defined nanopores with tailored properties, opening up new avenues for scientific exploration and technological advancements in diverse fields. The understanding and mastery of solid-state nanopore fabrication techniques contribute significantly to the development of novel sensors, DNA sequencing technologies and nanopore-based devices with applications in biotechnology, medical diagnostics and beyond (Edel, 2020).

8.4. PRINCIPLES OF NANOPORE MEASUREMENTS

The principles of nanopore measurements revolve around the fundamental interactions between analytes and the nanopore. When an analyte of interest passes through a nanopore, it induces measurable changes in electrical, optical, or force-based signals, allowing for its detection, characterization and even manipulation at the single-molecule level. These measurements can provide valuable information about the analyte's size, shape, charge, concentration and interactions with other molecules.

8.4.1. Ion Transport Through Nanopores

One of the key factors that make nanopores an excellent tool for measurements is their ability to facilitate the transport of ions. When a nanopore is immersed in an electrolyte solution, such as a saline buffer, it forms a conductive path for the flow of ions. The movement of ions through the nanopore is governed by the principles of electrostatics, diffusion and electrokinetics (Radenovic, 2019).

Electrostatic forces play a crucial role in ion transport through nanopores. The charged nature of the nanopore surface creates an electric field within the pore, influencing the movement of ions.

Positively charged ions (cations) are attracted to the negatively charged surface of the nanopore, while negatively charged ions (anions) experience repulsion. This electrostatic interaction results in the migration of ions toward or away from the nanopore, depending on their charge.

Diffusion, another important phenomenon, contributes to the movement of ions through nanopores. According to Fick's laws of diffusion, ions tend to move from regions of higher concentration to regions of lower concentration. As a result, ions diffuse through the nanopore, driven by concentration gradients, until equilibrium is reached (Edel, 2020).

Electrokinetic effects also come into play in nanopore measurements. These effects arise from the coupling between electric fields and fluid flow within the nanopore.

When an electric potential is applied across the nanopore, it generates an electric field that influences the motion of ions. This electric field, in turn, creates fluid flow within the nanopore, known as electroosmotic flow. Electroosmotic flow can enhance or hinder ion transport, depending on the direction and strength of the electric field.

8.4.2. Current Blockade Method

When a nanopore is exposed to an electrolyte solution and an electrical potential is applied across it, an ionic current flows through the pore. As molecules traverse the nanopore, they interrupt this ionic current, leading to a temporary reduction or "blockade" in the measured current known as Voltage driven current blockade. This change in current can be used to extract information about the properties of the translocating molecules, such as their size, shape, charge and interactions with the pore facilitated by the voltage driven current blockade.

Voltage driven current blockade is a fundamental technique employed in nanopore measurements. It relies on the application of an electric field across a nanopore, typically by applying a voltage difference across two electrolyte reservoirs containing the nanopore. This voltage drives the movement of ions through the nanopore and facilitates the detection of molecules passing through the pore (Radenovic, 2019).

When a molecule approaches the nanopore, it can partially or completely block the ionic current passing through the pore. This blockade effect, known as the ionic current signatures, leads to changes in the measured current, which can be recorded and analyzed to gather valuable information about the translocating molecule. By monitoring these current fluctuations, researchers can gain insights into various molecular properties, such as size, shape, charge and conformation.

Moreover, the detection and analysis of ionic current signatures play a central role in nanopore measurements. As different molecules interact with the nanopore, they induce characteristic changes in the measured current, resulting in distinct current signatures. These signatures serve as a fingerprint for the molecules passing through the nanopore, enabling their identification and characterization (Edel, 2020).

The ionic current signatures arise from the interaction between the molecules and the nanopore. Factors such as the size, shape, charge distribution and surface properties of the translocating molecule influence the magnitude and duration of the current blockade. By carefully analyzing the variations in the current, researchers can decipher valuable information about the properties of the molecules under investigation.

Moreover, the ionic current signatures can be further modified by altering experimental parameters such as the applied voltage, ionic strength of the electrolyte, pH and temperature. By systematically manipulating these variables, researchers can gain additional insights into the molecular behavior and probe specific interactions between the nanopore and the translocating molecules (Radenovic, 2019).

8.4.3. Translocation Events and Signal Analysis

Translocation events refer to the passage of molecules or particles through a nanopore, which is a tiny aperture typically on the order of a few nanometers in diameter. The nanopore can be a naturally occurring protein channel, such as alpha-hemolysin, or a synthetic pore engineered from solid-state materials, such as silicon nitride or graphene. When a molecule approaches the nanopore, it experiences a localized change in the electrical field, leading to alterations in the measured electrical signal.

The analysis of translocation events begins with the acquisition of raw nanopore signals, typically in the form of ionic current measurements. These measurements are obtained by applying a voltage across the nanopore and monitoring the resulting current as molecules translocate through the pore. The recorded signal contains valuable information about the size, shape and properties of the translocating molecules.

To extract meaningful information from the raw signal, signal analysis techniques are employed. Signal processing algorithms play a crucial role in filtering noise, resolving overlapping translocation events and characterizing the features of individual translocation events. These algorithms aim to enhance the signal-to-noise ratio and reveal the underlying patterns and characteristics of the translocating molecules (Edel, 2020). One common approach in signal analysis is event detection, which involves identifying and segmenting individual translocation events from the continuous signal. This is achieved by setting appropriate thresholds or employing statistical methods to distinguish the signal associated with translocation events from the background noise. Once the events are detected, their parameters, such as duration, amplitude and shape, can be analyzed to extract valuable information about the translocating molecules (Radenovic, 2019).

Moreover, machine learning techniques have been increasingly employed in nanopore signal analysis. By training algorithms on large datasets, machine learning models can learn to recognize specific patterns or features associated with different molecules or analytes.

This enables automated identification and classification of translocating molecules, opening up possibilities for high-throughput analysis and real-time monitoring of diverse samples.

However, despite the significant progress made in signal analysis for nanopore measurements, challenges still exist. One major challenge is the presence of noise in the measured signals, which can arise from various sources, including electronic noise, environmental fluctuations, or molecule-pore interactions. Developing robust denoising algorithms and improving experimental setups to minimize noise are ongoing areas of research (Edel, 2020).

Additionally, the interpretation of nanopore signals can be complex due to the diverse factors influencing the translocation process. Factors such as the size, charge and conformation of the translocating molecules, as well as the properties of the nanopore itself, can affect the measured signals. Thus, comprehensive models and theoretical frameworks are being developed to understand and predict the relationship between the observed signal and the underlying molecular properties.

8.5. APPLICATIONS OF NANOPORE MEASUREMENTS

The ability of nanopore-based measurements to analyze molecules at the single-molecule level in real time using nanopores has revolutionized many fields, offering new insights and capabilities that were previously unattainable. In this section, we will explore some of the key applications of nanopore measurements and discuss their significance and impact (Radenovic, 2019).

8.5.1. DNA Sequencing

Nanopore-based measurements have revolutionized DNA sequencing, offering a powerful and versatile approach to deciphering the genetic code. Traditional DNA sequencing techniques, such as Sanger sequencing and next-generation sequencing (NGS), have played pivotal roles in advancing genomics and personalized medicine. However, they suffer from limitations in terms of cost, time and sample preparation requirements. Nanopore sequencing, on the other hand, overcomes many of these obstacles and presents a promising alternative for high-throughput and real-time DNA analysis (Figure 8.4).

In nanopore-based DNA sequencing, a nanopore is a tiny hole, typically on the order of a few nanometers in diameter, embedded within a synthetic membrane. This nanopore acts as a single-molecule sensor that can detect individual DNA strands as they pass through it. The process begins by preparing a sample containing DNA molecules, which are then driven through the nanopore using an electric field. As the DNA traverses the nanopore, it causes disruptions in the ionic current flowing through the pore. These disruptions, known as "current blockades," can be measured and analyzed to infer the sequence of the DNA molecule (Radenovic, 2019).

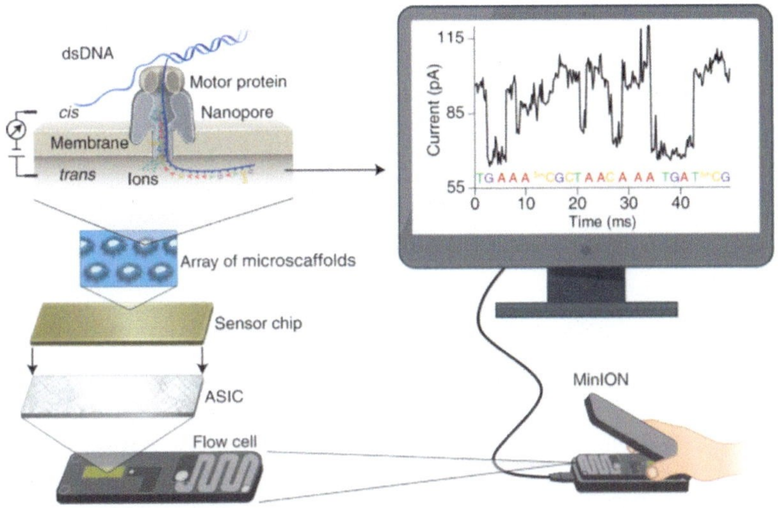

Figure 8.4. Principle of nanopore sequencing.

Source: Wang, Y., Zhao, Y., Bollas, A. et al. Nanopore sequencing technology, bioinformatics and applications. Nat Biotechnol 39, 1348–1365 (2021). https://doi.org/10.1038/s41587-021-01108-x

One of the key advantages of nanopore sequencing is its ability to generate long reads, which is crucial for resolving complex genomic regions, repetitive sequences and structural variations. Unlike NGS techniques that rely on fragmenting the DNA into small pieces, nanopore sequencing can analyze DNA molecules that are hundreds of kilobases in length. This capability greatly simplifies the assembly and annotation of genomes, leading to more accurate and comprehensive genomic information (Edel, 2020).

Furthermore, nanopore sequencing offers real-time analysis, allowing researchers to observe the sequencing process as it unfolds. This feature is particularly advantageous in situations where rapid detection of specific sequences or identification of pathogens is required, such as in infectious disease surveillance or outbreak investigations. Real-time sequencing also enables feedback-based experimental design, where subsequent analysis can be tailored based on the preliminary results, saving time and resources.

Another notable benefit of nanopore-based DNA sequencing is its versatility. Not only can it sequence DNA, but it can also analyze other types of nucleic acids, such as RNA. This opens up opportunities for transcriptomic studies, gene expression analysis and the detection of RNA modifications. Additionally, nanopore sequencing can provide insights into epigenetic modifications, such as DNA methylation patterns, which play a critical role in gene regulation and disease development (Edel, 2020).

Despite its tremendous potential, nanopore sequencing is not without challenges. One of the primary concerns is the relatively high error rates associated with the technology. The accuracy of nanopore sequencing is influenced by multiple factors, including pore size, DNA translocation speed and base-calling algorithms. Efforts are being made to improve the accuracy through advancements in pore engineering, signal processing techniques and bioinformatics algorithms (Radenovic, 2019).

Moreover, nanopore sequencing is still a developing field and standardization of protocols and data analysis methods is ongoing. Interlaboratory comparisons and benchmarking studies are essential to establish robust quality control measures and ensure consistent and reliable results across different platforms and laboratories. Collaboration between academic researchers, industry experts and regulatory bodies is crucial for advancing nanopore sequencing technologies and establishing their clinical utility.

8.5.2. Protein Analysis

Proteins play a crucial role in various biological processes and are involved in numerous diseases, making their analysis essential for understanding cellular functions and developing effective therapeutics. Traditionally, protein analysis has relied on techniques such as gel electrophoresis, mass spectrometry and fluorescence-based assays. While these methods have provided valuable insights, they often require complex sample preparation, suffer from limited throughput and are time-consuming.

Nanopore-based measurements offer a promising alternative for protein analysis due to their unique capabilities. In a nanopore setup, a nanometer-sized pore is embedded in a thin membrane, allowing the passage of individual molecules. By applying a voltage across the nanopore, researchers can detect changes in the ionic current as proteins translocate through the pore. This provides information about their size, structure and even specific interactions with ligands or other molecules.

One of the key advantages of nanopore measurements in protein analysis is their ability to analyze proteins in their native state, without the need for extensive sample preparation or labeling. This is particularly valuable as protein structure and function can be altered during purification and labeling processes. Nanopore measurements allow for the direct detection of proteins, preserving their natural conformation and activity (Edel, 2020).

Additionally, nanopore-based measurements offer high sensitivity and single-molecule resolution. Each protein translocation event produces a characteristic current signature, enabling the identification and analysis of individual proteins. This is especially beneficial for heterogeneous protein samples, where traditional ensemble methods may provide an average measurement but fail to capture the diversity within the population. With nanopore measurements, researchers can study protein variants, detect rare protein conformations and gain insights into dynamic processes that occur at the single-molecule level (Radenovic, 2019).

Furthermore, nanopore-based protein analysis can provide information about protein-ligand interactions. By modifying the nanopore surface or incorporating specific ligands within the pore, researchers can investigate the binding affinity, kinetics and thermodynamics of protein-ligand complexes. This opens up opportunities for drug discovery and personalized medicine, as nanopore measurements can assess the effectiveness of potential therapeutics and identify interactions that may be relevant to specific individuals or patient groups.

Despite the significant potential of nanopore-based protein analysis, several challenges remain. One major hurdle is the development of robust and reproducible nanopore fabrication techniques. The fabrication process needs to ensure consistent pore size, stability and control over surface properties to achieve accurate and reliable measurements. Additionally, the signal-to-noise ratio and resolution of nanopore measurements need to be optimized to detect and analyze proteins efficiently.

Interpreting the complex current signals generated by protein translocations also requires advanced data analysis methods. Machine learning algorithms and statistical models are being employed to decode the rich information contained in the nanopore current data, enabling the identification of specific protein features and structural characteristics.

8.5.3. Detection of Nanoparticles and Biomarkers

The detection of nanoparticles using nanopore measurements holds tremendous potential in a wide range of fields, including environmental monitoring, healthcare diagnostics and materials science. Nanoparticles, with sizes typically ranging from 1 to 100 nanometers, exhibit unique properties and behaviors compared to bulk materials. Nanopore-based detection methods allow for the real-time analysis of individual nanoparticles, providing insights into their size, shape, composition and surface charge. This information is invaluable for understanding nanoparticle behavior, assessing their potential impact on human health and the environment and optimizing their applications in various industries (Edel, 2020).

One of the key advantages of nanopore measurements in nanoparticle detection is their ability to distinguish between different particle types based on their characteristic electrical signatures. As particles pass through the nanopore, changes in the ionic current can be recorded and analyzed. Each nanoparticle type generates a distinct current signal due to differences in size, shape and surface properties. By comparing these signals with a reference database, it is possible to accurately identify and classify nanoparticles, even in complex mixtures (Radenovic, 2019).

Furthermore, nanopore-based measurements offer exceptional sensitivity, enabling the detection of nanoparticles at ultra-low concentrations. Traditional detection methods often rely on bulk measurements, which can limit sensitivity and specificity. In contrast, nanopore measurements provide a single-particle level analysis, enhancing the detection limit and enabling the identification of rare or low-abundance nanoparticles. This sensitivity is particularly critical in the field of biomarker detection. Biomarkers are specific molecules or particles that serve as indicators of biological processes, disease states, or response to therapies. They play a crucial role in early disease diagnosis, monitoring treatment efficacy and predicting patient outcomes. Nanopore-based measurements offer a promising avenue for biomarker detection due to their ability to analyze individual biomolecules with high precision.

In nanopore-based biomarker detection, a nanopore is functionalized with recognition elements such as antibodies, aptamers or enzymes that selectively bind to the target biomarker. As the biomarker molecules interact with the functionalized nanopore, changes in the ionic current are detected, providing a direct readout of biomarker presence and concentration. This approach eliminates the need for complex sample preparation and amplification steps, reducing the overall analysis time and improving the reliability of results.

The versatility of nanopore-based measurements allows for the detection of a wide range of biomarkers, including proteins, nucleic acids and small molecules. For example, in cancer diagnostics, specific protein biomarkers associated with different types of tumors can be identified using nanopore measurements, facilitating early detection and personalized treatment strategies. Similarly, in infectious disease monitoring, nanopore-based measurements can detect nucleic acid biomarkers from pathogens, enabling rapid and accurate diagnosis (Radenovic, 2019).

Moreover, the portability and miniaturization potential of nanopore devices open up possibilities for point-of-care applications. By integrating nanopore measurements into portable devices or lab-on-a-chip systems, healthcare professionals can perform on-site biomarker detection, reducing the reliance on centralized laboratories and improving patient care in resource-limited settings (Edel, 2020).

8.6. ADVANCEMENTS AND FUTURE DIRECTIONS IN NANOPORE TECHNOLOGY

Over the years, significant progress has been made in the development of nanopore sensing devices. Initially, biological nanopores, such as the α-hemolysin protein pore, were primarily used for nanopore measurements. However, the limitations associated with biological nanopores, such as low stability and difficult control of their properties, prompted researchers to explore alternative options.

One of the major advancements in nanopore sensing devices is the development of solid-state nanopores. These nanopores are typically created by drilling a tiny hole in a thin membrane, usually made of materials like silicon nitride or graphene. Solid-state nanopores offer several advantages over their biological counterparts, including improved stability, precise control over pore size and shape and the possibility of integrating them with other nanostructures (Radenovic, 2019).

Additionally, researchers have been investigating the use of hybrid nanopores, which combine the advantages of both biological and solid-state nanopores. These hybrid nanopores leverage the unique properties of biological pores while taking advantage of the stability and tunability of solid-state nanopores. By incorporating biological channels into solid-state membranes, researchers have been able to achieve better control over the sensing process and enhance the overall performance of nanopore devices.

Accurate and reliable signal analysis, on the other hand, is crucial for extracting meaningful information from nanopore measurements. Advancements in signal analysis techniques have played a vital role in improving the sensitivity and specificity of nanopore sensing.

Signal processing algorithms, such as Hidden Markov Models (HMMs) and machine learning approaches, have been widely employed to analyze the complex and noisy signals obtained from nanopore measurements. These techniques enable the identification and discrimination of different analytes passing through the nanopore based on their unique signal patterns. By training the algorithms on large datasets, researchers have been able to improve the accuracy of nanopore measurements, paving the way for applications in genomics, proteomics and single-molecule analysis (Edel, 2020).

Moreover, real-time signal analysis has gained significant attention in nanopore technology. The ability to process signals in real-time allows for rapid analysis and enables the development of high-throughput nanopore devices. This advancement has led to applications in DNA sequencing, where nanopore devices are capable of reading the DNA sequence as it passes through the pore in real-time, revolutionizing the field of genomics.

Furthermore, to expand the capabilities of nanopore technology, researchers have been exploring its integration with other complementary technologies. One notable integration is the combination of nanopore sensing with nanofluidics. Nanofluidic channels can be integrated with nanopores to precisely control the movement and delivery of analytes to the sensing region. This integration enables efficient sample preparation, enhances analyte capture rates and improves overall measurement sensitivity.

Subsequently, the integration of nanopores with electronic circuits has shown great promise. By incorporating nanopores into electronic devices, such as Field-Effect Transistors (FETs), researchers have been able to directly measure electrical signals while simultaneously monitoring analyte interactions. This approach not only provides real-time electrical readouts but also offers the possibility of label-free detection and high-resolution sensing (Radenovic, 2019).

However, the advancements in nanopore technology have paved the way for its commercialization and adoption in various industries. One of the most prominent applications is DNA sequencing. Nanopore-based sequencing devices offer rapid and cost-effective DNA analysis, with the potential to revolutionize personalized medicine, diagnostics and forensics.

Moreover, nanopore technology has found applications in drug discovery and development. Nanopores can be used to study the interactions between drugs and their targets, providing valuable insights into drug efficacy, kinetics and off-target effects. This enables more efficient drug screening processes and the development of personalized therapies.

In the field of environmental monitoring, nanopore-based sensors show promise for the detection of pollutants and contaminants in water, air and soil. Their high sensitivity, portability and real-time analysis capabilities make them suitable for on-site monitoring applications (Edel, 2020).

Therefore, the advancements and future directions in nanopore technology are driving its widespread adoption across various industries. The development of novel nanopore sensing devices, enhanced signal analysis techniques, integration with other technologies and its commercialization in applications like DNA sequencing, drug discovery and environmental monitoring are opening up new opportunities and revolutionizing fields that require rapid and sensitive analysis. Continued research and innovation in nanopore technology will undoubtedly lead to further breakthroughs and expanded applications in the coming years (Radenovic, 2019).

8.7. CHALLENGES AND LIMITATIONS OF NANOPORE MEASUREMENTS

However, nanopore-based measurements have gained significant attention in recent years due to their potential for high-resolution sensing and analysis of various biomolecules, including DNA, RNA, proteins and small molecules, they face several challenges and limitations that need to be addressed for their widespread adoption and application in various fields. Noise and signal-to-noise ratio represent a major challenge in nanopore measurements. Noise can arise from various sources, including thermal fluctuations, electronic circuitry and ionic fluctuations in the nanopore system. This noise can significantly degrade the quality of the measured signals and affect the accuracy of the obtained data. Achieving a high signal-to-noise ratio is crucial for reliable and precise nanopore measurements. Researchers have employed various strategies to mitigate noise-related challenges, such as implementing advanced signal processing algorithms, optimizing experimental conditions and developing improved nanopore fabrication techniques. However, achieving a consistently high signal-to-noise ratio across different nanopore platforms and experimental setups remains a significant hurdle.

Resolution and accuracy are two key parameters that determine the quality and reliability of nanopore measurements. Resolution refers to the ability of the nanopore system to distinguish between different analytes or molecules with similar physical properties. It is influenced by several factors, including the size and shape of the nanopore, the applied voltage and the ionic conditions. Achieving high resolution in nanopore measurements requires careful optimization of these parameters and precise control over experimental conditions. Additionally, accuracy, which refers to the degree of agreement between the measured values and the true values of the analyte properties, is also critical. Accurate nanopore measurements rely on calibration standards, well-defined experimental protocols and accurate data analysis techniques. Enhancing both resolution and accuracy is an ongoing research focus in the field of nanopore-based sensing (Radenovic, 2019).

Scalability and throughput present significant challenges in nanopore measurements, particularly when dealing with large-scale or high-throughput applications. While nanopore systems offer the potential for single-molecule analysis, the ability to scale up these measurements to handle high volumes of samples simultaneously is crucial for their practical use in fields such as genomics, proteomics and clinical diagnostics. Achieving high throughput requires the development of multiplexing strategies, integration of nanopore arrays and parallelization of measurement platforms. Furthermore, challenges related to data acquisition, storage and analysis also need to be addressed to ensure efficient and scalable nanopore measurements.

Another critical challenge in nanopore measurements is device reproducibility and stability. Nanopore fabrication involves complex processes and achieving consistent and reproducible nanopore characteristics across different devices and batches is nontrivial. Variations in pore size, shape and surface properties can lead to inconsistent and unreliable measurements. Additionally, nanopore devices may exhibit temporal instability, where their performance changes over time due to factors such as pore fouling, drift in pore size, or electrode degradation. These issues pose significant challenges to the reproducibility and long-term stability of nanopore measurements. Efforts are underway to improve fabrication techniques, enhance device robustness and develop reliable quality control protocols to address these challenges (Edel, 2020).

While nanopore-based measurements hold immense potential for high-resolution and label-free analysis of biomolecules, they also face several challenges and limitations that must be overcome for their broader adoption. Noise and signal-to-noise ratio, resolution and accuracy, scalability and throughput, as well as device reproducibility and stability, are among the key challenges that researchers are actively addressing. By developing advanced signal processing techniques, optimizing experimental conditions and improving fabrication processes, it is possible to overcome these challenges and unlock the full potential of nanopore-based measurements in various scientific and technological domains (Radenovic, 2019).

Overall, nanopore technology has emerged as a powerful and versatile tool holding the future of various scientific and technological applications. Nanopores can be broadly classified into biological nanopores and solid-state nanopores. Biological nanopores, such as protein channels, offer high specificity and selectivity but are limited in their controllability and stability. On the other hand, solid-state nanopores, including synthetic nanopores and carbon nanotubes, provide greater controllability and stability, making them suitable for a wide range of applications.

Fabrication techniques for nanopores depend on the type of nanopore being used. Biological nanopores are typically obtained from natural sources or can be engineered using molecular biology techniques. Solid-state nanopores can be fabricated using various methods such as electron beam lithography, focused ion beam milling and chemical etching, among others (Edel, 2020).

The principles of nanopore measurements involve the ion transport through nanopores, which leads to current blockade. By analyzing the translocation events and the resulting signal, valuable information about biomolecules, such as DNA sequencing, protein analysis and the detection of nanoparticles and biomarkers, can be obtained. These applications have revolutionized fields like genomics, proteomics and diagnostics, enabling faster and more affordable analyzes with high accuracy.

Despite the remarkable advancements in nanopore technology, there are still challenges and limitations that need to be addressed. One of the major challenges is enhancing the accuracy and speed of DNA sequencing, especially in long-read sequencing. Improvements in signal-to-noise ratios, reducing artifacts and minimizing the impact of pore clogging are ongoing areas of research (Edel, 2020). Future directions in nanopore technology include the development of new materials with enhanced properties, such as improved sensitivity and selectivity. Integration of nanopores with other analytical techniques and miniaturization of devices are also being explored to enable portable and point-of-care applications. Additionally, advancements in data analysis and computational tools will be crucial for harnessing the full potential of nanopore measurements (Radenovic, 2019).

In conclusion, nanopore technology holds immense promise for a wide range of scientific and technological applications. With continued research and development, overcoming the current challenges and limitations, nanopore-based measurements are poised to revolutionize fields such as personalized medicine, environmental monitoring and biotechnology, paving the way for new discoveries and innovations.

CHAPTER 9

Fluorimetry and Its Applications

Contents

9.1. INTRODUCTION

Fluorimetry measures and analyzes the fluorescence of substances in response to a certain wavelength of light. This requires identifying the photons produced by a sample when it goes from excited to reduced energy. Counting spectroscopy measures radiation from an item. Radiation is emitted as electrons travel from excited to ground. Fluorescence spectroscopy studies fluorescence. Spectrofluorimetry is fluorimetry (Figure 9.1).

It requires a beam of light, generally ultraviolet, to excite the electrons inside molecules of specific substances to generate light, usually visible but sometimes not. Absorption spectroscopy also works. Single-molecule fluorescence spectroscopy detects intensity fluctuations using fluorophores or mixes. Fluorometers measure fluorescence.

Energy levels are molecule's numerous states. Fluorescence spectroscopy studies electronic and vibrational states. The species under research often has a ground and agitated electrical state of interest. Each electrical state has several vibrational states. Fluorescence requires photon absorption to stimulate a species from its ground electronic state to one of its excited electronic state vibrational modes. The stimulated molecule loses vibrational energy via collisions with other molecules until it reaches the lowest vibrational state possible from its excited electrical state (Groves, 2020).

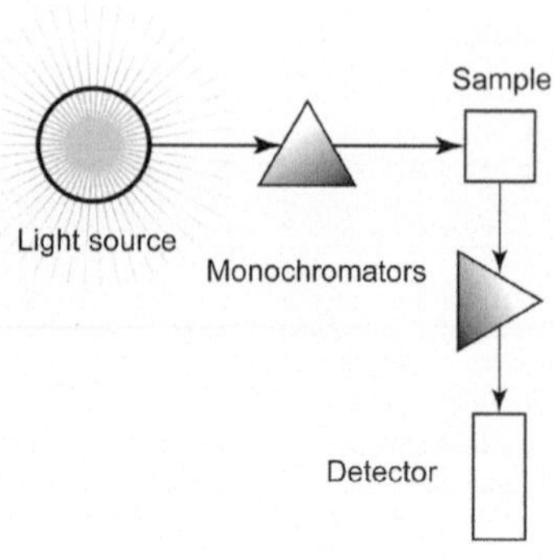

Figure 9.1. Schematic illustration of the components of a fluorimeter.

Source: Carlsson, Nils. (2013). Spectroscopic Studies of Biomolecules Confined in Self-Assembled Nanostructures. Scientific Figure on ResearchGate. Available from: https://www.researchgate.net/figure/Schematic-illustration-of-the-components-of-a-fluorimeter_fig4_256148181 (accessed on 22 September 2023)

A photon is released when the molecule falls to one of the ground electronic state's vibrational states. Because molecules in the ground state can reach numerous vibrational modes, the emitted photons will have a wide range of energy and frequency. Fluorescence spectroscopy may be used to study light frequencies and intensities to determine vibrational level structure. Atomic species lack vibrational energy levels, therefore photons discharged typically have the same wavelength as radiation going in. "Resonance fluorescence" occurs in molecular and atomic fluorescence. It involves re-emitting absorbed photons. Fluorescence excitation measurements have a fixed detection wavelength and variable excitation wavelengths within a region of interest, whereas fluorescence (emission) measurements have fixed excitation and variable detection wavelengths. An emission map captures and combines excitation wavelength emission spectra. This three-dimensional surface data set depicts emission intensity vs excitation and emission wavelengths, like a contour map (Abraham, 2018).

9.2. INSTRUMENTATION

Filter and spectrofluorometers are instruments. Spectrofluorometers utilize diffraction grating monochromators, whereas filter fluorometers use filters. Both methods filter and monochromatic excitation light before it reaches the sample. Some sample molecules emit light, while others absorb light. Fluorescent light radiates. A detector is commonly positioned at a 90-degree angle to the incoming light beam to reduce incident light transmission or reflection. Before reaching the detector, fluorescent light passes via a second monochromator (Groves, 2020).

Lasers, LEDs, Xenon arcs and Mercury-vapor lamps can excite. A monochromator or filter is not needed for excitation because a laser creates high-irradiance light at 0.01 nm. This method cannot significantly vary laser wavelength. Mercury vapor lamps provide peak wavelength light. However, Xenon arc emission spectra are continuous, with a reasonably constant intensity in the 300–800 nm area and enough irradiance to allow studies down to 200 nm.

Fluorimeters employ filters or monochromators. Monochromators transmit adjustable wavelength and tolerance light. The most popular monochromator employs a diffraction grating to contact collimated light and exit at a wavelength-dependent angle. Adjusting the monochromator selects transmission wavelengths. Anisotropy measurements require two polarization filters—one after the excitation monochromator or filter and one before the emission monochromator or filter (Abraham, 2018). As noted, fluorescence is measured at 90° to the excitation light. Instead of positioning the sensor at 180° to the transmitted excitation light, this architecture avoids interference. Poor monochromators let extraneous light and other wavelengths through. The ideal monochromator would only transmit light within the defined range and have good wavelength-independent transmission. Stray light is merely sample light when measured at 90°. Compared to 180° geometry, this increases the signal-to-noise ratio and decreases the detection limit by 10,000. Fluorescence from the front of opaque or hazy items is another approach. Detectors may have many channels. Multichannel detectors can simultaneously detect all wavelength intensities, unlike single-channel detectors. Emission monochromators and filters are unnecessary (Groves, 2020). The most flexible fluorimeters use twin monochromators and a continuous excitation light source to record excitation and fluorescence spectra. While the emission monochromator scans fluorescence spectra, the excitation light should be at a wavelength with substantial absorption. The excitation monochromator measures excitation spectra using the emission filter or monochromator wavelength. Because fluorescence intensity is inversely linked to absorption, excitation and absorption spectra overlap.

The UV-Visible spectroscopy equipment is identical to the fluorescence spectroscopy equipment. It possesses the following

- Three types of lamps are used for lighting: Tungsten, mercury vapor and Xenon arc.
- Monochromators: These devices aid in dividing light bands.
- Sample cells: The majority of the sample cells are quadrilateral. The material is glass.
- Detectors: This photomultiplier tube aids in the detection of low-intensity radiation.
- Filter fluorometers and spectrofluorometers are two examples of fluorimetric apparatus.
- The spectrofluorometer is the most sophisticated and effective of them all.
- Light *Source:* The source must produce consistent radiation that is strong enough to cover the whole spectrum.
- Xenon lamp: Spectrofluorometers frequently utilize this. Here, the light is created by applying high pressure to ionized Xenon

gas while it is being electrically conductive. It emits powerful radiation across a large area and produces a dazzling glow. So, the majority of substances may be examined (Abraham, 2018).

- The mercury vapor in this lamp is exposed to a very high pressure of eight atmospheres. Over 350 nm, it radiates strongly. This vapor also emits another radiation at 254 nm when the pressure is low. However, this is favored in low-cost fluorimeters of the filter kind.
- Tungsten light: This lamp emits low-intensity radiation. When the excitation radiation is visible, it is employed (Groves, 2020).
- Monochromators and filters: The light source generates polychromatic light. They must thus be transformed into monochromatic light using filters or monochromators. The light radiation has a bandpass of about 30 nm since the filter's resolution capability is lower. As a result, they are cheap and are employed in filter fluorometers. The monochromators, in contrast, offer a very high-resolution capability with a very narrow bandpass of just 0.1 nm. However, they are pricey, which is why spectrofluorometers.
- Exemplary cells -The sample is held in the fluorometer cuvettes, which are tiny tube-like objects used for analysis. Fused glass with color correction is utilized to make them. The cell's sidewalls are polished on every side. The sample holder's route measures 10mm in length.

Photo-multiplier tubes are used as detectors in science. These are accurate and effective for low-intensity radiation.

9.3. ADVANTAGES FLUORESCENCE SPECTROSCOPY

- Sensitivity: Fluorimetry, which can evaluate samples in minute amounts, is a very effective spectroscopic technique. It is decided on sample concentrations as low as g and ng/ml.
- Precision: This fluorimetry spectroscopy has a high degree of result precision. A margin of 1% or more is feasible.
- Specificity: UV-VIS spectroscopy only uses the substance's excitation wavelength. Wavelengths of both excitation and emission are typical in fluorescence spectroscopy. The emission

wavelengths of two substances would differ if they had the same excitation wavelength. Thus, a compound's specificity is improved (Groves, 2020).

9.4. DISADVANTAGES OF FLUORIMETRY

Despite all of this method's advantages, there are some drawbacks. These are associated with fluorescent brightness.

- Fluorescent intensity can be impacted by pH variations and the presence of oxygen.
- Because UV radiation has a high intensity, it can cause chemical changes in samples when it is absorbed.
- Not every material can be made luminous. Therefore, this approach cannot be used to examine all chemicals.
- The presence of electronegative halides and heavy metals can alter the fluorescence intensity.
- Fluorescence spectroscopy's fundamentals -The transition of electrons from the singlet ground state to the singlet excited state is caused by the absorption of UV or visible light. This state releases energy in the form of UV or visible light since it is not stable and it then decays back to the singlet ground state. As the fluorophore decays from the singlet electronic excited states to an acceptable vibrational level in the electronic ground state, fluorescence emission takes place (Abraham, 2018).

9.5. ANALYSIS OF DATA

Fluorescence intensity frequently decreases with fluorophore concentration at low doses. Device-independent "standard" spectra are harder to produce than UV/visible spectra. Since various factors alter spectra, changes are needed to generate "true" or machine-independent spectra. This section classifies aberrations as sample- or instrument-related. First, check instrument distortion. The light source's strength and wavelength vary between researchers. No lightbulb has a uniform spectrum intensity. After the excitation monochromator or filter, a beam splitter can direct light to a reference detector (Groves, 2020).

A secondary reference detector should follow the excitation monochromator or filter. The mechanism controls how much fluorescence the detector detects. The detector's quantum efficiency, or proportion of seen photons, fluctuates with wavelength, detector type and detector degradation.

The radiation optics and sample container (cuvette or cell) are two further aspects to consider. Most UV, visible and near-infrared studies require quartz cuvettes. Both circumstances demand materials with minimal wavelength absorption. Quartz transmits between 200 and 2500 nm and better-grade quartz may transmit up to 3500 nm, which helps as other material's absorption may conceal the sample's fluorescence (Abraham, 2018). The difficult procedure of adjusting each instrumental component to produce a "standard" spectrum is only utilized when necessary. Calculating the quantum yield or the wavelength with the strongest emission is an example. The sample also distorts. Thus, sample components must be examined. Photodecomposition reduces fluorescence intensity progressively. Consider light diffusion too. Unlike Raman scattering, Rayleigh scattering keeps scattered light at the same wavelength as incoming light. The excitation light-induced virtual electronic state causes Raman scattering. From this hypothetical situation, molecules may decompress to a vibrational level other than the vibrational ground state. Fluorescence spectra always feature a peak below the excitation wavenumber, like 3600 cm-1 in water (Groves, 2020). Also, consider the internal filter, Resorption is one. Reabsorption occurs when the fluorophore absorbs a macromolecule or part of it at its wavelength. The fluorophore's photons may be reabsorbed. The fluorophore is one of the numerous absorbing molecules that add inner filtering. Thus, excitation light intensity fluctuates throughout the solution. Thus, fluorophores visible to the detecting system get just a tiny percentage of excitation light. Inner filter effects modify the fluorescent light output spectrum and intensity (Abraham, 2018).

9.6. TRYPTOPHAN FLUORESCENCE

The fluorescence of a folded protein is a composite of the fluorescence from several aromatic residues. Tyrosine, phenylalanine and tryptophan residues contribute very little to the intrinsic fluorescence emissions of an unfolded protein, but disulfide connections absorb a lot of light in this wavelength range. Tryptophan has a maximum absorption wavelength of 280 nm and a solvatochromic emission peak that ranges between 300 and 350 nm depending on the polarity of its surroundings. Protein fluorescence can therefore be used to assess a protein's structural state.

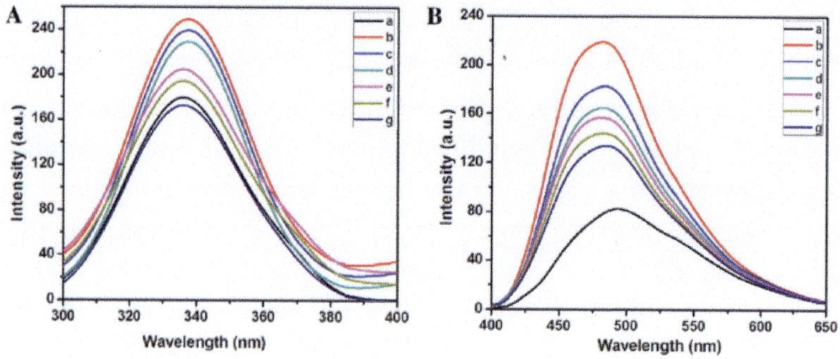

Figure 9.2. (A) Intrinsic tryptophan fluorescence and (B) ANS fluorescence emission spectra of (a) native b-lg, heat exposed (75°C), (b) b-lg, (c) b-lg:FA (1:1), (d) b-lg:FA (1:3), (e) b-lg:FA (1:4), (f) b-lg:FA (1:5) and (g) b-lg:FA (1:10). b-lg concentrations throughout all emission experiments were kept at 0.25 mg/mL.

Source: Pal, Sampa & Maity, Sanhita & Sardar, et al (2020). Antioxidant ferulic acid prevents the aggregation of bovine β-lactoglobulin in vitro. Journal of Chemical Sciences. Scientific Figure on ResearchGate. Available from: https://www.researchgate.net/figure/A-Intrinsic-tryptophan-fluorescence-and-B-ANS-fluorescence-emission-spectra-of-a_fig1_343594276 (accessed on 22 September 2023)

Additionally, the tryptophan fluorescence is significantly influenced by the presence of nearby residues (e.g., nearby protonated groups like Asp or Glu may lessen Trp fluorescence). Tryptophan and the other fluorescent amino acids may also transmit energy, which might affect the research, especially when using the Forster acidic approach.

Additionally, tryptophan is a very rare amino acid and many proteins only include one or a few tryptophan residues. Tryptophan fluorescence allows for extremely precise monitoring of the conformational state of each tryptophan residue. Intrinsic probes work better since the protein is not changed (Figure 9.2). The use of intrinsic fluorescence to assess protein structure is often limited to situations with a small number (or maybe only one) of tryptophan residues since each tryptophan residue is exposed to a different local environment, resulting in distinct emission spectra (Groves, 2020).

To ascertain the surroundings of the tryptophan, an essential amino acid with intrinsic fluorescence can be employed. Using denaturants, surfactants, or other amphiphilic substances may change the tryptophan microenvironment during research. A red-shifted emission spectrum will be seen, for instance, if a protein with a single tryptophan in its "hydrophobic" core denatures with rising temperature. This happened as a result of tryptophan being exposed to water instead of the hydrophobic core of a protein. If the tryptophan is incorporated in the surfactant vesicle or micelle, adding a surfactant to a protein and exposing it to an aqueous solvent causes a blue-shifted emission spectrum. Proteins lacking in tryptophan may be linked to a fluorophore. The tryptophan emission spectrum dominates over the weaker tyrosine and phenylalanine fluorescence when fluorescence is activated at 295 nm (Abraham, 2018).

9.7. APPLICATIONS OF FLUORIMETRY

Fluorescence spectroscopy is used in a variety of scientific fields, including biochemical, medicinal and chemical research, to examine organic substances. It has also been proposed that it can be utilized to differentiate between benign and malignant skin tumors. AFS methods, like CVAFS, may be used for a range of studies or measurements of a component present in air, water, or other media. CVAFS is used to identify heavy metals like mercury. More information about redirecting light using fluorescent materials may be found in fluorescent solar collectors. Another method for bringing fluorescence spectroscopy down to the microscopic level is microfluorimetry. Fluorescence detectors are used in combination with HPLC in analytical chemistry. When studying water, fluorescence spectroscopy may be used to detect organic contaminants and assess the water's quality. Even the detection of bacterial contamination in water has been made possible by recent developments in computer science and machine learning (Groves, 2020).

- ***Analytical chemistry***

Analytical chemistry routinely uses fluorimetry to assess the quantitative and qualitative characteristics of diverse substances. For locating and measuring fluorescent pigments, markers and probes in intricate biological materials, it is incredibly useful. To make it easier to detect and monitor molecules of interest, fluorescent markers and labels are typically attached to such molecules. Fluorescence spectroscopy is a useful tool for analytical chemists. All molecules absorb electromagnetic radiation, even though the vast majority do not emit light. To separate fluorescent components from complicated mixtures, fluorescence spectroscopy is a useful approach. Currently, fluorescence spectroscopy is used in almost all academic disciplines, from biology to physics. The identification of DNA, RNA, proteins, enzymes and other macromolecules has been accomplished using it. Metals with dangerous quantities below the nanometer range have been found. Toxic metals including mercury, arsenic, chromium, nickel, cobalt and vanadium are first complexed with the proper substances to produce luminous agents. Fluorescence intensity and concentration have an inverse relationship; hence it is possible to estimate the quantity of metal in the complex. Another method that is becoming more prominent in fluorescence-based chemical research is fluorescence quenching. In this case, the natural fluorescence of the reagent is extinguished by the addition of a metallic salt solution. The amount of metal is determined by the amount of fluorescence quenching (Abraham, 2018).

- ***Environmental monitoring***

Monitoring and evaluating the environment requires the use of fluorimetry. It is used to find and measure organic chemicals, heavy metals and pesticides in samples of soil, water and air. Fluorimetry sensors may be made to bind preferentially with particular analytes, making them very sensitive and selective for environmental monitoring. The creation of biosensors with a wide range of applications has been made easier by the rise in interest in nanomaterials with advanced optoelectronic properties. Due to the necessity to quickly identify toxins in water sources, fluorescent biosensors have become more common in the environmental sector. These tools are simple to use and very sensitive. Nanomaterials including carbon dots (CD), quantum dots (QD), gold nanoparticles, nanoclusters, graphene (G), graphene oxide (GO), transition metal dialcogens (TMDC) and organometallic frameworks (MOF) have been used to create fluorescent detecting systems. They frequently have the Forster resonance energy transfer (FRET) quenching idea as their base. Even in real-world samples, the target pollutant may be specifically detected using nanomaterials bioconjugated with molecules like antibodies, enzymes, or aptamers. Basic sanitation is harmed by heavy metals released by mining and industry, so-called emergent pollutants released from a range of sources, the majority of which are urban and pesticides used to boost agricultural productivity. Numerous illnesses, including Hodgkin's lymphoma, Parkinson's disease, cancer, issues during pregnancy and endocrine anomalies have been related to pesticide exposure. The endocrine system may also be impacted by acute exposure to several substances that have been recognized as emerging contaminants for some time. Personal care items, home cleaners, sunscreens, cosmetic additives, raw materials for the plastics sector and pharmaceutical leftovers are a few examples.

These compounds are extremely difficult to identify using existing analytical methods due to their extremely low quantities. Meanwhile, heavy metals are found in water sources, particularly in developing or impoverished countries with a strong extractivist inclination. Heavy metals that cannot be eliminated or broken down can be dissolved or precipitated by physicochemical agents or the medium's environment, such as pH or temperature. Some of them create soluble chemicals that are spread across ecosystems and carried there before they enter the food chain. All of the pollutants mentioned above constitute a concern to human health and ecosystems due to their propensity for bioaccumulation, which has long-term repercussions, in addition to the acute impacts of some of them (Groves, 2020). Due to their user-friendliness, the potential for cheap component prices and capacity for real-time quantification, biosensors have quickly arisen as a complement to existing analytical methods for the identification of pollutants of interest in waterways. In analytical devices known as biosensors, a mediator element is in touch with a transducer element, which is in charge of translating the interaction into a quantifiable signal. Numerous biomediators and transducers are available, but optical and electrochemical biosensors hold the most promise for precise and accurate contamination detection. The inherent fluorescence of nanomaterials, particular metal cations and particular biological processes are required for the improved fluorescent biosensors to operate without labeling (Abraham, 2018).

- ***Biochemical and biomedical research***

Fluorimetry is widely employed in these fields of study. It makes it possible to examine biological processes, protein folding, enzymatic reactions and molecular interactions. One popular method in molecular biology to analyze protein-protein interactions and conformational changes inside biomolecules is fluorescence resonance energy transfer (FRET). Fluorescence spectroscopy applications to protein and peptide analysis are influenced by three main variables: the signal's dynamic character, its confined nature and its redundancy. Although these characteristics may make it more difficult to comprehend the results of the experiment, they may also be used to gather specific structural and dynamic data. When compared to other spectroscopic methods, fluorescence is more widely used due to its accessibility and ease of basic data collecting and processing. However, this simplicity does not seem to detract from its benefits. Fluorescence correlation spectroscopy (FCS) is a potent method for quantifying molecular dynamics and has found extensive use in a variety of disciplines, including chemistry, biomedicine and biophysics. FCS can quantitatively assess the concentration, diffusion coefficient and interaction of the molecules in vitro or in vivo by time-correlating the fluorescence variations caused by molecules diffusing via focused light. The basis of life events in biological systems is molecular dynamics. To identify molecular dynamics, fluorescence correlation spectroscopy (FCS) analyzes the intensity variation that biomolecules produce as they move in and out of a focused light. With FCS, precise measurements may be made of the local concentration, hydrodynamic radius, diffusion coefficient, interactions between various proteins, etc. FCS provides a wider measurable time range, from ns to s and a greater sensitivity that can reach a single-molecule level when compared to other dynamics-oriented techniques (Groves, 2020).

To evaluate the binding of the fluorescent dye EtBr with DNA. Originally suggested FCS in 1972. Ever since FCS was initially used in conjunction with confocal microscopy in the 1990s, it has been a useful tool for studying molecular dynamics. Confocal microscopy offers a very constrained observation volume of less than 0.5 fl, which considerably improves FCS's signal-to-noise ratio (SNR).

FCS is always confounded by several issues when studying various dynamic processes in living biological systems, including cell mobility, fluorophore photobleaching, anomalous dynamics in sub-diffraction zones, etc. Over the past 20 years, several advancements have been made to increase FCS's potential to overcome these challenges. To mention a few, dual-color FCCS was suggested as a method of determining how various compounds interact. FCS can investigate dynamics at several sites thanks to space-multiplexing techniques. Parallel/perpendicular line-scanning FCS overcomes the photo-bleaching caused by the slow diffusion of molecules on the membranes by quickly scanning the focus on the membrane or perpendicular to the membrane, while pair correlation function (pCF) based FCS provides information on the directional motion of molecules along a certain direction, i.e., between a pair of points. Additionally, the perpendicular line-scanning FCS eliminates membrane movement-induced false fluorescence signal fluctuation. We can detect anomalous diffusion in the nanoscale areas thanks to the interaction of FCS with stimulated emission depletion (STED), which lowers the detection volume. To efficiently minimize the observation volume along the z-axis and obtain molecular information from thousands of pixels simultaneously, single-plane illumination microscopy-based FCS (SPIM-FCS) integrates light-sheet illumination with FCS. By examining the fluctuation of unlabeled particles in a highly labeled environment, Inverse-FCS calculates the particle volume independent of diffusion dynamics. Additionally, there are data analysis extensions that can yield molecular brightness that is not possible with FCS, such as photon counting histogram (PCH) and fluorescence intensity distribution analysis (FIDA). These improvements broaden the scope of traditional FCS in several ways and enable the measurement of biomolecular dynamics and interactions in intricate biological settings (Groves, 2020).

- ***Drug discovery and development***

Fluorimetry is a crucial component of the procedures involved in drug discovery and development. To find promising therapeutic candidates, it is used for high-throughput screening of chemical libraries. To track binding interactions, evaluate medication efficacy and track enzymatic activities, fluorophores can be added to drug molecules or binding assays. The biophysical approach known as differential scanning fluorimetry (DSF) has been used for a variety of purposes throughout the years, including the detection of protein folding states and the discovery of ligands that bind to the target protein. Modern drug development efforts are driven by biophysics, which enables quick and high-throughput data capture to screen through vast chemical libraries in search of novel bioactive compounds. The temperature shift test, also known as differential scanning fluorimetry (DSF), is a crucial tool in this biophysical arsenal. DSF is a commonly used biophysical approach that is affordable, parallelizable, useful and accessible for tracking both protein folding state and heat stability. By gradually heating the protein in a controlled setting, it offers an effective technique for examining protein unfolding. The process of protein denaturation may be seen by detecting the variations in fluorescence emission that occur in response to temperature increases (Abraham, 2018).

The method has found several successful applications and has been applied in many ways in recent years since protein thermal stability is affected by changes in sample behavior through complex formation with even weakly binding ligands. For many years, it has predominantly been used as a drug discovery technique to find potential lead compounds for a variety of target proteins. The identification of ideal conditions for storage, assay screening and crystallization is a significant use of DSF in protein buffer optimization. The best buffer components for each distinct protein may be found by screening sparse matrix settings, including several buffer systems that cover a wide range of pH, additions and salt concentrations. In the past, it has been demonstrated that doing so increased the success rates of protein crystallization. More recently, DSF has also been used to address the issue of sample preparation. Two publications show how appropriate screening methods can be used to identify and optimize sample refolding buffers, enabling significantly less expensive access to the protein sample amounts needed to support high-throughput screening campaigns. Last but not least, a fairly recent advancement has demonstrated that DSF is capable of offering accurate data in challenging situations, such as unpurified chemical processes. This is a noteworthy finding since chemical entity manufacturing and purification constitute a significant bottleneck in every screening program (Groves, 2020).

Although the DSF approach has become a crucial biophysical tool in drug development due to its robustness and broad application in sample preparation and screening, it is necessary to be aware of its limitations. This is especially true when planning a screening campaign since such a campaign should include orthogonal screening choices that are not subject to the same restrictions to reduce both false positives and false negatives. DSF is a reliable biophysical method for examining the stability of proteins in a given environment, whether it be under certain buffer conditions or when they are (partially) saturated with target ligands. The thermodynamic factor for protein unfolding no matter if the target protein was in a pure form, in lysate, cells, or even tissues, Tm is monitored as the principal sign to validate stability changes of the target protein. In particular, recently developed label-free nano DSF approaches eliminate the need for dyes, enabling the same approach to be applied to research on membrane proteins while simultaneously addressing issues brought on by the interaction between dye and the hydrophobic surface of proteins, or the detergent additives used and interactions between the dye and other molecules in a screen. With the help of low sample consumption and high throughput, the DSF technique has been used to characterize the thermal properties of many proteins over the nearly 20 years since it first appeared. This has made DSF suitable for both screening large ligand libraries and optimizing buffer components in crystallization. Although there have been many successful cases of ligand-binding validation reported in the literature, it is still important to be aware that this correlation typically occurs for similarly structured compounds within a series and persistently pursuing fragment hits based on significant thermal shifts may lead to further optimization being misled. Also keep in mind that ligands can interact with target proteins in both their folded and unfolded states and a decrease in melting temperature does not exclude binding to the native state. DSF is sensitive to all stages along a binding pathway, making it difficult to use to assess the affinity of molecules toward mobile protein receptors. In contrast to titration-based techniques like ITC,

MST and SPR, which depend on different serial concentrations of ligands and end-point measurements, DSF is sensitive to all stages along a binding pathway. The robustness and adaptability of DSF to solve a variety of issues across such a broad range of sample types, however, should guarantee that it remains a key technique in contemporary drug development (Groves, 2020).

- ***Clinical diagnostics***

Fluorimetry has a wide range of clinical diagnostic uses. The detection and quantification of biomarkers, hormones, antibodies and genetic material. Fluorescent probes and pigments are used in DNA sequencing, flow cytometry and immunoassays to help diagnose diseases and track the health of patients. Chemical devices that use fluorescent probes are necessary in the disciplines of chemical biology and medicine. Finding better and more effective ways to diagnose various diseases is becoming more and more crucial as new medical technologies are created. These technologies might range from precise molecular tools for early cancer and genetic disease identification to pathogen detection technology. The bulk of existing diagnostic techniques take a long time and cost a lot of money to complete. They also need trained workers, specialized equipment and specialist facilities. The development of biosensors offers a much-needed substitute for current diagnostic techniques that is accurate and free of logistical hassles. These gadgets should ideally be useful, durable, small, portable, economical and able to deliver information that non-expert users may readily understand. Biosensors can be utilized for imaging, health monitoring and drug effectiveness evaluation in addition to diagnosis. A key component of clinical diagnostics is the use of biomarkers to pinpoint specific illnesses and during the past ten years, the number of publications in the field of biosensors for clinical applications has drastically expanded. "Any substance, structure, or process that can be measured in the body or its products and that influences or predicts the incidence of an outcome or disease" is how the World Health Organization defines biomarkers. These include of nucleic acids, hormones, proteins, certain metabolites and proteins. Depending on the kind, they might offer information on therapeutic response, prognosis, prevention, or diagnosis (Abraham, 2018).

The effectiveness of therapy is predicted by predictive biomarkers, sometimes called therapeutic response-related biomarkers. Preventive biomarkers assess the risk of infection and susceptibility to illness. Diagnostic biomarkers are connected to disease detection. For routine analysis in hospitals or even less specialized healthcare facilities, a screening panel for the most prevalent biomarkers, including PSA, the BRCA1 and BRCA2 genes for breast cancer and the EGFR gene for lung cancer, could be combined and implemented. Systems for providing healthcare would benefit from this. This technique is capable of detecting a broad variety of illnesses, including bacterial infections and cancer. Optical imaging technologies are critical to both basic science and clinical medicine. One of the numerous reasons in favor of its broad usage is the fact that it is non-intrusive and that biological processes may be studied with great temporal and spatial accuracy. Chemists have focused on creating chemical instruments that can distinguish between different biological species and keep track of key physiological processes in order to improve the accuracy, sensitivity and range of optical imaging applications. Fluorescent probes are the current name for these chemical instruments. Numerous uses of fluorescent probes, including immunofluorescence labeling, live-cell imaging, medication administration (theranostics) and fluorescence-guided surgery have proven to be beneficial. The intricacy of biological processes and the restricted tissue penetration depth of light, however, provide several challenges. High cell/tissue specificity, selectivity for the target biomarker in the presence of competing molecules, photostability, superior biocompatibility and long-wavelength excitation are qualities of an ideal fluorescence probe. The capacity to recognize intracellular creatures and keep track of bodily functions has aided in the creation of new techniques for illness detection and improved our comprehension of biology. A fluorophore, a receptor or reactive site and a communication pathway often make up a fluorescence probe. This makes sure that when the instrument interacts with the target analyte, a quantifiable result is seen. The fluorophore and recognition unit of fluorescent probes created for the same biomarker may differ dramatically as a result of continual improvements in synthetic methods and photophysical sciences. This is demonstrated by the creation of fluorescent probes for the detection of the reactive molecules peroxynitrite-boronate and -ketoamide. However, the observed fluorescence-based responses either involve a reduction in fluorescence emission intensity (fluorescence quenching) or an increase in fluorescence emission intensity (fluorescence enhancement), or a change in the wavelength of the fluorescence emission, which may result in ratiometric sensing. Fluorescence probes continue to be designed using the same core principles, though. Fluorescent probes often fall into one of two groups: receptor-based or reaction-based. Even though the emphasis in this modality is on an absorption shift, it is crucial to highlight that the creation of photoacoustic (PA) imaging agents follows the same design principles. To identify ions, such as metal cations and anions, through coordination or complexation interactions, fluorescent probes based on receptors are widely utilized. With reaction-based fluorescence probes, which are used to detect reactive species and enzymes, a biomarker contact causes a chemical reaction that changes the fluorescent probe's structure and emissive characteristics. This review has been separated into parts based on the many kinds of biological species and offers examples of recent research for the target species to help readers understand the applications of receptor-based and reaction-based systems in detail (Groves, 2020).

- ***Material's science***

Fluorescence-exhibiting compounds are characterized and examined via fluorimetry. It is used to look into the makeup, structure and characteristics of materials including nanoparticles, polymers and semiconductors. The electrical structure and behavior of these compounds may be extensively studied using fluorescence spectroscopy. Characterization in materials science refers to the laborious process of analyzing and measuring a material's structure and characteristics. Without it, engineered materials would not be understood scientifically. The field of materials research depends on it. Other definitions allow the phrase to refer to any technique of materials analysis, including macroscopic procedures like mechanical testing, thermal analysis and density computation. Some definitions restrict the term's use to methods that analyze the microscopic structure and characteristics of materials. The size of observable structures in materials characterization ranges from centimeters, where coarse grain patterns in metals may be seen, to angstroms, where individual atoms and chemical bonds can be distinguished (Abraham, 2018).

While certain characterization techniques, like basic optical microscopy, have been in use for many years, new methods and procedures are constantly being created. Particularly in the 20th century, the development of the electron microscope and secondary ion mass spectrometry transformed the discipline by making it possible for researchers to photograph and analyze structures and compositions on far smaller sizes than was previously feasible. Our comprehension of the reasons why various materials display distinctive features and behaviors has significantly advanced as a result (Groves, 2020).

AFM's maximum resolution for examining certain materials has significantly improved during the past three decades. A characterization method called microscopy explores and maps a material's surface and subsurface structure. These technologies may gather information about a material's structure at various length scales using photons, electrons, ions, or physical cantilever sensors. Microscopy methods include SEM laser microscopy, TEM scanning electron microscopy and transmission electron microscopy. SPM is sometimes referred to as atomic force microscopy, STM scanning tunneling microscopy, XRT X-ray telescope and field ion microscopy (FIM) and Radiation Diffraction Topography. A collection of techniques known as spectroscopy is used to ascertain a material's chemical composition, composition variation, crystal structure and photoelectric characteristics. Examples of spectroscopic methods include the following:

- Ultraviolet-visible (UV-Vis) optical spectroscopy
- FTIR is the abbreviation for Fourier transform infrared spectroscopy.
- The thermoluminescence symbol is TL, whereas the photoluminescence symbol is PL.

Food and beverage analysis

Fluorimetry is utilized for quality assurance and analysis in the food and beverage sector. It may be used to find and measure pollutants, residues and additions in food items. It may be used, for instance, to gauge the amount of aflatoxins, a particular class of mycotoxin, in food samples. For the identification of either inherent or foreign chemicals, the use of fluorescence spectroscopy in foods is extremely important. In addition to proteins, peptides, amino acids, amines, vitamins and inorganic species, antioxidant capacity assessments have been added among intrinsic compounds.

Exogenous chemical applications include the fluorometric techniques mostly discussed for polycyclic cyclic hydrocarbons, mycotoxins, phycotoxins, pesticides, microbial contamination and veterinary medicine residues, which were researched in earlier iterations of this page. A new section has been included due to the presence of acrylamide and the significance of its correct regulation in heat-processed meals.

Food and beverage safety is becoming a major concern for everyone's health, especially about food quality and hygiene. Consuming food is a daily requirement for everyone in the world for health, development and illness prevention. Therefore, it's crucial to make sure that consumers only receive reputable, safe food goods. However, in practice, customers find it challenging to select legal food goods because of low food quality, poor cleanliness, adulteration, impurity, outdated food and other problems (Abraham, 2018).

One of the main consumer concerns in developing nations is food adulteration since it directly affects people's health. Due to lower manufacturing costs and more market demand than supply, particularly in light of the most recent COVID-19 pandemic circumstances, food adulteration occurs. As a result of the food being altered or combined with other legal or illegal substances, consumers are unable to detect contaminated food using their senses and must instead rely on laboratory testing. To ensure food safety and public health, it is crucial and should be a top priority to establish food and beverage evaluations. For the assessment of food and beverages, several methodological techniques and technologies have been developed, including image, aroma, taste, electromagnetic sensing and others. The detection of poor food quality, falsified food and adulteration are major issues for consumers today. Given the intrinsic complexity of food items, the majority of instrumental approaches used for evaluating quality and authenticity take a lot of time, money and work to complete. As a result, there has been an increase in demand for quick, easy and accurate analytical methods for evaluating food quality features. Absorption spectroscopy in the UV-Vis (ultraviolet and visible) area is one of these methods, utilized for the quantitative and qualitative analysis of sample chemicals. This method has previously been applied in several food science and food processing research domains due to its simplicity and dependability (Groves, 2020).

- ***Pharmaceutical analysis***

For medication formulation and quality control, fluorimetry is widely employed in pharmaceutical analysis. It may be used to examine the stability and degradation of medications under various situations as well as the concentration and purity of active pharmaceutical ingredients (APIs). The word "pharmaceutical analysis" is wide and has several definitions. It is a collection of procedures used to identify, determine, separate, purify and elucidate the chemical structure of a specific substance utilized in the creation of pharmaceutical goods. Drug metabolites, pharmaceutical excipients, pharmaceutical contaminants, or active pharmaceutical substances are often the components to which pharmaceutical analysis is applied. Biological samples, completed pharmaceutical products, impurities, toxins and pharmaceutical raw materials are often used as samples in pharmaceutical analysis. Various analytical techniques can be used to do pharmaceutical analysis. Human health underwent a revolution as a result of the discovery of drugs. These medications would only fulfill their purpose if they were supplied in the right quantity and were free of contaminants. Different chemical and instrumental techniques that are used in drug assessment have been developed over time to make medications function as intended. These medications must be recognized and quantitated since contaminants can form in them at different points during production, transit and storage, making administration dangerous. Analytical tools and techniques are crucial in this regard. Spectrophotometric approaches based on inadvertent UV absorption and chemical reactions are another significant set of techniques that are frequently seen in pharmacopeias (Gorog, 1995). The quantitative assessment of a material's reflection or transmission characteristics concerning wavelength is called spectrophotometry (Abraham, 2018).

These approaches have the benefit of requiring little work and time. These techniques have remarkable accuracy. Over the past several years, there has been a sharp rise in the usage of UV-Vis spectrophotometry, particularly when it comes to the examination of medicinal dosage forms. The pharmaceutical industry is always searching for delicate analytical methods that use tiny samples. One method that achieves great sensitivity without sacrificing specificity or precision is fluorescence spectrometry. The primary goal of pharmaceutical medications is to benefit people by keeping them healthy and preventing sickness. The medication must be devoid of impurities or other disturbances that might damage humans for it to fulfill its intended function (Groves, 2020).

- ***Conclusion***

Fluorimetry has a wide range of uses and these are only a few of them. The approach is a valuable tool for scientific study, industrial applications and medical diagnostics due to its sensitivity, selectivity and non-destructive nature. A photon source is used to excite sample molecules in the sensitive optical emission method known as fluorescence spectroscopy. By measuring the intensity of such emission, it is possible to identify the molecules that relax through radiant emission.

When a colorimetric approach is insufficiently sensitive or selective to identify a chemical, fluorimetry is often utilized. Immunoassays and the chemistry of bioluminescence are among the significant applications for determining chemical and inorganic molecules that are addressed. In this work, certain fluorimetric uses are also covered. It appears simple to do fluorescence studies at first appearance. But many things can mess up the data and make the findings incorrect. The potential for sample contamination or signal contamination due to stray or dispersed light must always be considered. All experiments need the collection of emission spectra and the analysis of blank samples (Abraham, 2018).

Principle of fluorescence spectroscopy

Absorption of UV or visible radiation causes transition of electrons from singlet ground state to the singlet excited state. As this state is not stable, it emits energy in the form of UV or visible radiation and returns to singlet ground state. Fluorescence emission occurs as the fluorophore decay from the singlet electronic excited states to an allowable vibrational level in the electronic ground state.

I. **Factors affecting fluorescence**

1. **Conjugation**: Molecule must have unsaturation i.e., it must have π electrons so that UV-Vis radiation can be absorbed. If there is no absorption of radiation, there will not be fluorescence.
2. **Rigidity of structures:** Rigid structures will produce more fluorescence, while flexible structure will produce less fluorescence.
3. **Nature of substituent groups:** Electron donating groups like amino, hydroxyl groups enhance fluorescence activity. Electron withdrawing groups like Nitro, carboxylic group reduce fluorescence. Groups like SO_3H or on NH_4 + have no effect on fluorescence intensity.
4. **Effect of temperature:** Increase in temperature leads to increase in collisions of molecules and decrease in fluorescence intensity while decrease in temperature leads to decrease in collisions of molecules and increased fluorescence intensity.
5. **Viscosity:** Increase in viscosity leads to decreased collisions of molecules which will enhance fluorescence intensity while decrease in viscosity causes increased collisions of molecules which cause decreased fluorescence intensity.
6. **Oxygen:** Decreases the Fluorescence intensity in two ways: Oxidizes fluorescence substances to non-fluorescence substances. It quenches fluorescence because of paramagnetic properties.
7. **Effect of pH:**

a. **Aniline:** Neutral or alkaline medium shows visible fluorescence while acidic conditions give fluorescence in UV region only.

b. **Phenols :** Acidic conditions do not give fluorescence while alkaline conditions gives good fluorescence.

II. Effect of concentration on fluorescence intensity:

Fluorescence intensity = Q x Ia

where, Q = Fluorescence efficiency; Ia = Intensity of absorbed light Since emission is proportional to absorption; Ia has to be known.

To compute the concentration, the formula A = εcL can be used, where A is the absorbance measured at the peak wavelength, ε is the extinction co-efficient, c is the concentration, and L represents the path length. The associated extinction coefficients at the peak wavelength are εCy3 = 150,000 L/mol-cm, εCy5 = 250,000 L/mol-cm, εCy5.

CHAPTER 10

The Future of Biophysical Chemistry

Contents

10.1. INTRODUCTION

As we approach the future, biophysical chemistry has a lot of potential and exciting opportunities. This helps us understand how tiny things like atoms and molecules work together with big things like living things. It helps us figure out the rules that control how living things and chemicals interact. Every day, new ways of doing things, like tools and ways of thinking, are pushing the boundaries of this field, taking us into areas of knowledge that we haven't explored before (Gouverneur, 2012)

Biophysical chemistry has the potential to drastically change our comprehension of life on a tiny scale in the future. New progress in understanding the structure of cells and organelles has allowed us to see and study their complex structures and how they interact in a clearer way than ever before.

Nanotechnology is a field that can control really tiny particles. It can be used to deliver medicine directly to specific parts of the body, make accurate medical tests and create new materials for the body. Synthetic biology is a growing area that mixes engineering and biology to create artificial life forms and materials inspired by nature (Landes, 2017).

Additionally, combining biophysical chemistry with quantum mechanics has led to the emergence of quantum biology. This field investigates how quantum phenomena impact biological systems.

The research on quantum effects in enzymes, photosynthesis and sensory systems could completely change our understanding of how living things use and control energy.

Besides these areas of science, biophysical chemistry has significant impacts on medicine and treatments. Being able to understand how molecules in our body interact with each other and create medicines based on this knowledge can help personalize medical treatments. At the same time, finding new ways to treat cancer and edit genes brings us closer to fighting diseases that have been harming humans for a long time.

As we start our journey into the future of biophysical chemistry, we have to deal with both difficulties and chances. We need to pay attention to important things like ethics, analyzing data, using computers to model things and working together with different fields of study. However, every challenge has the possibility of big discoveries that will change the way we do science and medicine (Gouverneur, 2012).

10.2. ADVANCEMENTS IN STRUCTURAL BIOLOGY AND BIOMOLECULAR INTERACTIONS

In the world of science, different fields of study are coming together to make big discoveries about how living things work. Scientists have discovered hidden information within the tiny world of molecules by combining ideas from different areas of study. Specifically, studying the structure of living things and how different molecules interact with each other has given us deep knowledge about the basic processes of living organisms. Using advanced methods, scientists can now see the detailed structure of biological molecules and understand how they work together in cells. These discoveries have helped us understand more about tiny particles and have also started the groundwork for new ways of treating illnesses (Landes, 2017).

- ***Cryo-electron microscopy and single-particle analysis***

One of the biggest improvements in the study of how living things are structured is the development and use of a method called cryo-electron microscopy (cryo-EM) and single-particle analysis techniques. This new method has changed how we can see and understand the shapes of big molecules in three dimensions. We can now see them more clearly and with more detail than ever before. In the past, scientists used X-ray crystallography and nuclear magnetic resonance (NMR) spectroscopy as the main ways to find out the shapes of big biological molecules. But these methods had problems when it came to big and complicated molecules, membrane proteins and big complexes. Cryo-EM can solve these problems and offers a strong alternative.

In cryo-electron microscopy, samples are quickly frozen in a special kind of ice called vitreous ice. This freezing process helps to maintain the original structures of the samples and reduces harm from radiation. After that, the samples are seen through a special microscope that uses electrons. A bunch of pictures of tiny particles are taken. These pictures are used by a computer to create a 3D map of the molecule (Gouverneur, 2012).

Single-particle analysis is an important part of cryo-EM. It involves organizing and aligning individual particles in the images to create a detailed structure. This method helps scientists study the shapes of materials that can change or vary. It allows them to see different ways that molecules can move and understand how they work. The use of cryo-EM and single-particle analysis has had a big effect on the study of structures in biology. It has made it easier to find structures that were hard to find before, like big groups of molecules, viruses and proteins in cells. Scientists using cryo-EM have achieved very detailed images of complex molecules, which has given them a better understanding of how these molecules work.

In addition, cryo-EM has made it easier for researchers all over the world to study the structure of biological molecules. The invention of cryo-EM machines that can be used for business, better automation and easy-to-use computer programs have made the process of collecting and analyzing data much easier. This has made cryo-EM available to more people, allowing more scientists to study the details of structural biology (Landes, 2017).

In the future, cryo-EM and single-particle analysis will keep advancing the field of structural biology. New technologies like better detectors, improved image processing methods and combining cryo-EM with other techniques have the potential to improve image quality and help us study even more complicated systems. These new advancements will definitely help us learn more about how biological molecules work and how they relate to each other. This will also help us find new ways to treat diseases on a very detailed level (Gouverneur, 2012).

- ***Hybrid methods***

Scientists are using various techniques to learn more about the shapes of biomolecules. By combining different methods, these mix-and-match approaches can assist us in gaining a better understanding of how big molecules are structured and how they move within living organisms. Hybrid methods often use various scientific techniques such as freezing electron microscopy, X-ray crystallography, magnetic resonance spectroscopy and small-angle X-ray scattering together to achieve improved outcomes. Each way of studying the molecule has its own advantages and provides unique information about it. By bringing together different sets of information, scientists can develop a clearer understanding of how complicated biological systems are put together and work (Landes, 2017). For instance, scientists have used both cryo-EM and X-ray crystallography to figure out detailed structures of big molecular groupings. Cryo-EM gives us a basic rough structure, which we then use as a starting point to create a more detailed model using information obtained from X-ray crystallography. This combination of methods has been very helpful in understanding the shapes of complicated molecular machines, like the ribosome and the spliceosome. In another example, combining NMR spectroscopy and cryo-EM has been helpful in studying how biomolecules move and change shape. NMR spectroscopy helps understand the different shapes and movements of molecules in liquids, while cryo-EM shows us the fixed structures of molecules. By using these methods together, scientists can understand and explain the shape changes and movement of biomolecules that are important for their function (Gouverneur, 2012).

Additionally, hybrid methods also include the combination of real-life experiments with computer modeling and simulations. Combining real-world data with computer simulations and modeling helps scientists improve structures, study different shapes and movements and examine the energy and interactions of biomolecules in great detail. Using hybrid methods can help us learn more about how biological systems work, because they are both complicated and always changing. By using different methods, researchers can overcome limitations and get a better understanding of biomolecular structures and interactions. These combined methods will keep enabling new findings in fields like creating medicines, improving proteins and understanding how diseases work. As technology and computer tools get better, hybrid methods will become more important in understanding the secrets of life at the tiny molecular level (Landes, 2017).

- ***Membrane protein structure determination***

Proteins in cell membranes are very important for many biological processes, like sending signals between cells, moving molecules across cell membranes and helping with chemical reactions. But figuring out how membrane proteins are made has always been hard because they are not attracted to water and interact in complicated ways with layers of fat. In the past few years, new ways of studying membrane proteins have helped us learn more about these important biological molecules (Gouverneur, 2012).

One of the new methods that has changed how we study the structure of membrane proteins is cryo-electron microscopy (cryo-EM). Cryo-EM helps us see membrane proteins in their natural fatty surroundings, helping us understand their shapes and how they work. Scientists have used a combination of cryo-EM, computer image processing and modeling techniques to figure out detailed structures of difficult membrane proteins. This has helped them understand how these proteins are structured and how they work.

Another effective way to find out the structure of membrane proteins is by using X-ray crystallography. While it is more difficult to crystallize membrane proteins compared to soluble proteins, there have been advancements in techniques like lipidic cubic phase (LCP) crystallization and using antibody fragments or nanobodies to improve the chances of success. X-ray crystallography has helped scientists see the detailed structure of different membrane proteins at the atomic level. This has made it possible to study how these proteins interact with other molecules, like drugs and how they can change shape.

Nuclear magnetic resonance (NMR) spectroscopy is a helpful method for studying how membrane proteins are built and how they move around in a liquid. NMR is a helpful tool that can give us important information about how proteins and lipids interact with each other, how they group together and how their shape may change. Improvements in preparing samples, using isotopic labeling and using powerful NMR instruments have made it easier to study membrane proteins. This allows scientists to figure out the structure of these proteins and understand how they work (Gouverneur, 2012).

Computers and simulations have been very important in finding out how membrane proteins are structured. Molecular dynamics simulations, along with real-world data, can improve and confirm membrane protein structures, study their movements and give us clues about how they work. These computer methods are very useful for studying how membrane proteins behave in lipid layers, which imitate the inside of a cell. Finding out the structure of membrane proteins is extremely important because it helps us understand how they work in living organisms. It also has important implications for creating new drugs and developing them. Knowing how membrane proteins are shaped helps create better medicines to target these proteins. This means the drugs can be more effective and have fewer side effects. As technology gets better and new ways of doing things come up, the study of how proteins are shaped is going to make even more progress. Using different methods like cryo-EM, X-ray crystallography, NMR spectroscopy and computational modeling can help us understand membrane proteins better and their importance in health and disease (Gouverneur, 2012).

- ***Unraveling protein folding and dynamics***

Protein folding is when a chain of amino acids transforms into its natural, three-dimensional structure. It is an important part of studying proteins. It is important to study how proteins fold and move in order to understand how they work and why they sometimes become unstable or cause diseases like Alzheimer's, Parkinson's and cystic fibrosis. In the past few years, scientists have made a lot of progress in understanding how proteins fold and how they change over time.

New and different ways of studying proteins, like NMR spectroscopy, single-molecule fluorescence and hydrogen-deuterium exchange have given us important information about how proteins fold. NMR spectroscopy helps us understand the structure and behavior of proteins by studying different stages of folding, temporary states and looking at specific parts of the protein's structure. Single-molecule fluorescence techniques help scientists see how proteins fold and unfold in real-time. This helps us understand how proteins change shape and the steps they take to do so. The hydrogen-deuterium exchange studies help us understand how fast hydrogen atoms move in different parts of a protein. This gives us information about when the protein folds or unfolds and how stable different parts of the protein are.

In addition, improvements in computer techniques have been very important in understanding how proteins fold and move. Molecular dynamics (MD) simulations use classical mechanics to simulate how proteins fold and study their movement at a very small scale. MD simulations help us understand how proteins fold by studying how individual atoms move and interact with each other. This gives us information about the different ways proteins can change shape and the energy involved in the folding process. The use of both testing in the lab and computer simulations has helped researchers understand how proteins fold in a way that is really impressive. For example, experiments can give us data to check and improve computer models. Computational methods can help come up with ideas to plan experiments and understand results better (Gouverneur, 2012).

New research has shown that protein folding is not a straightforward process and can vary widely. There are many different ways that proteins can fold and there are also stages or steps in between. Furthermore, scientists now understand that the way proteins move and change shape is very important for how they work.

Methods like time-resolved spectroscopy and single-molecule force spectroscopy have helped understand how proteins move and work. New discoveries about how proteins fold and move have many important implications. They can help in creating plans for changing proteins and making medicines that target misshaped proteins. By understanding why proteins fold incorrectly, scientists can find ways to treat or lessen the impact of diseases caused by this.

As technology gets better, using both experiments and computer simulation will help us learn more about how proteins fold and move. Combining detailed imaging techniques, advanced methods for analyzing molecules and complex computer algorithms will help us understand how proteins fold and move. This will lead to important discoveries in scientific research and the creation of new treatments.

- ***Advances in NMR spectroscopy and molecular modeling***

NMR spectroscopy helps us understand how atoms are connected and how molecules behave and move at very small scales. Now, with the help of new technologies like NMR spectroscopy and computer modelling, we can understand and study complex biological systems better than before. Making magnets that are more powerful is a big step forward in NMR testing. It is easier to find and study bigger molecules and smaller signals using NMR in stronger magnetic fields. NMR is now better for studying bigger proteins, mixtures of proteins, DNA and proteins in cell walls (Gouverneur, 2012).

Another important improvement is the creation of experiments using multidimensional NMR. By using different dimensions, like 3D or 4D experiments, scientists can learn more about the structure and motion of biomolecules. These experiments help us find out which amino acids are involved in interactions, measure how far apart they are from each other and understand how the shape of the molecules change and move. In addition, new ways of labeling have been developed for NMR spectroscopy. Isotope labeling techniques, like marking certain parts of proteins or DNA, help scientists study specific areas in more detail. This special labeling helps to study the structure and interactions of complex biological systems.

Molecular modeling, combined with NMR spectroscopy, has become an essential tool for comprehending and explaining experimental NMR data. Computational methods, like computer simulations and sampling techniques, can create models that support and confirm experimental NMR data. These models help us understand how biomolecules move and change shape, as well as the energy involved. They help us understand experimental data better and create more precise structural models.

Moreover, improvements in computing speed and algorithm improvement have sped up the area of studying and simulating molecules. Better force fields and scoring models help make better predictions about how molecules are shaped and interact with each other. Improved methods for studying the shapes of molecules and measuring how well they bind together have been developed to better understand how molecules behave. These methods help us explore different ways molecules can arrange themselves and predict how strongly they stick together. NMR spectroscopy and molecular modeling are very helpful in studying how proteins and drugs interact, how proteins fold and move and how their structure is determined. These new developments are very important for finding new drugs. They help us find potential targets for drugs and design treatments that work better and are more specific (Gouverneur, 2012).

To sum up, recent progress in a technique called NMR spectroscopy and computer modeling of molecules has greatly improved how we can examine complicated living things. The sensitivity, experiments in different dimensions, new labeling techniques and advanced computer algorithms have made NMR more useful and allowed us to learn more about the shape, movement and interactions of biomolecules. These new advancements are expanding our knowledge and creating opportunities for more scientific exploration and the creation of new healing methods.

10.3. NANOTECHNOLOGY AND BIOPHYSICAL CHEMISTRY

Nanotechnology and biophysical chemistry are two fields that come together and work together. This collaboration creates lots of chances to make big scientific progress and discoveries. Nanotechnology is the science of working with really tiny particles. It combines ideas from physics, chemistry and engineering to make small things that can do useful tasks. Biophysical chemistry is a field that uses ideas from biology, physics and chemistry to study the physical and chemical qualities of living things at a very small scale.

The combination of nanotechnology and biophysical chemistry has resulted in significant advancements in many fields. One important use is in the area of medicine where drugs are delivered to the body. Tiny particles called nanoparticles can be made to wrap around drugs.

This helps the drugs to dissolve better, last longer and go directly to certain cells or tissues. These tiny carriers can go through obstacles in our bodies and deliver medicine in a careful way. This makes the medicine work better and causes fewer problems for patients. Nanotechnology can be used to make special medicine that can be customized for each person based on their own characteristics (Gouverneur, 2012).

Another way these two fields overlap is by creating very small sensors for studying living things. Tiny sensors called nanosensors are able to identify and measure certain molecules or substances in our bodies. They are very sensitive and accurate in their detection.

By combining tiny materials with things that can recognize biological elements, small sensors can find signs of diseases, keep track of harmful substances in the environment and measure how our bodies are doing in real-time. Nanotechnology and biophysical chemistry working together can quickly and accurately help diagnose and track biological processes.

In addition, people are very interested in creating materials and devices that imitate nature. They are using nanotechnology and biophysical chemistry to do this. Researchers are able to make fake body parts and devices that work better and do more by copying how real bodies are built and work. Nanoscale engineering allows us to control the characteristics of materials, like how they feel on the surface, how strong they are and how well they work with living things.

This helps us make biomaterials that are very similar to natural biological systems. These materials and devices that imitate nature have the potential to help create new body tissues, treat diseases and develop improved sensors and devices that can be put inside the body.

Nanotechnology is also important in improving how we see things in biophysical chemistry. Tiny imaging techniques, like powerful microscopes and special imaging tools, can show us very detailed pictures of small things. These techniques can help us see and understand biological structures and processes better than ever before.

These imaging methods help us study how cells work, how molecules interact and how biomolecules behave. This helps us understand how living things work at a very small level. Combining nanotechnology and biophysical chemistry has great potential for solving problems in healthcare, environmental monitoring and scientific research. As tiny technology gets better and better, combining it with the study of how living things work will definitely lead to important discoveries that will change the future of science and technology.

10.4. QUANTUM BIOLOGY AND BIOPHYSICAL CHEMISTRY

Quantum biology is a new area of study that looks at how quantum physics and biophysical chemistry affect living things. It investigates how quantum phenomena play a part in biological systems.

In simple terms, biophysical chemistry is a field that studies how things work in living organisms. It uses basic ideas from physics and chemistry to understand these processes. But new discoveries show that quantum effects are important in many biological processes, which surprises scientists and allows for new research opportunities (Gouverneur, 2012). In quantum biology, scientists study how things like superposition, entanglement and tunneling affect the behavior of tiny living things, like molecules in our body, chemical reactions that happen inside us, how plants convert sunlight into energy and our ability to sense things. For instance, very small effects in how molecules react inside a living organism can make the reactions happen quicker and more effectively. This helps the organism's body processes to work incredibly accurately. In photosynthesis, scientists believe that quantum coherence helps plants to efficiently collect and use light energy to make food. Scientists are creating special sensors that can study and measure how things work at the tiny quantum level in living things. This will help us understand more about how biological systems function.

The combination of quantum biology and biophysical chemistry has great potential for helping us learn more about how life works. By using quantum principles, scientists can better understand how biological processes work and find new ways to use this knowledge in fields like biotechnology, medicine and energy. Studying how quantum physics connects to biology through biophysical chemistry can change the way we understand life. It highlights how quantum physics and biological systems are connected and challenges traditional scientific boundaries. As scientists study quantum biology more, it could help us make big discoveries in medicine, energy and computers and change the way we think about science and technology (Gouverneur, 2012).

10.5. BIOPHYSICAL CHEMISTRY IN MEDICINE AND THERAPEUTICS

Biophysical chemistry is very important in medicine and therapeutics. It helps us understand things better and come up with new ideas. It helps make drugs better and predicts how they interact with target molecules by using computer simulations and molecular docking. Biophysical techniques can be used to focus on problematic proteins that are not structurally organized and are involved in diseases such as cancer and neurodegenerative disorders. This helps scientists create targeted treatments that are tailored for these specific proteins. Biophysical methods help with personalized medicine by combining information from genes, proteins and molecules to find markers that can diagnose, predict outcomes and choose treatments for diseases.

Furthermore, biophysical methods help in treating cancer by using specific drug delivery systems and radiation therapy, which reduce harm to healthy tissues. Research on how proteins interact, how drugs stop working and how cancer cells live in their surroundings help us create new treatments and ways to treat cancer. Furthermore, biophysical chemistry is important in gene editing tools like CRISPR-Cas9, making gene therapy treatments more effective and safer. In general, studying the chemistry of living things helps make progress in finding new drugs, developing personalized treatments, fighting cancer and editing genes. This impacts the field of medicine and makes patients healthier.

In summary, the future of biophysical chemistry looks very promising and has a lot of potential. As scientists keep learning more, new technologies in biology and medicine will help us understand life better. In simple words: Biophysical chemistry in the future can make important discoveries and shape scientific progress by combining different areas of study and using new technology (Gouverneur, 2012).

Bibliography

1. Adamson, A. (2012). *A Textbook of Physical Chemistry*. Elsevier.
2. Albanese, A., Tang, P. S., & Chan, W. C. (2012). The effect of nanoparticle size, shape and surface chemistry on biological systems. *Annual Review of Biomedical Engineering, 14*, 1–16.
3. Allen, J. (2012). *Principles of Physical Sedimentology*. Springer Science & Business Media.
4. Arosio, P., Vendruscolo, M., Dobson, C. M., & Knowles, T. P. (2014). Chemical kinetics for drug discovery to combat protein aggregation diseases. *Trends in Pharmacological Sciences, 35*(3), 127–135.
5. Atkins, P., & De Paula, J. (2011). *Physical Chemistry for the Life Sciences*. Oxford University Press, USA.
6. Atkins, P., Atkins, P. W., & de Paula, J. (2014). *Atkins' Physical Chemistry*. Oxford university press.
7. Borgnakke, C. (2013). *Fundamentals of Thermodynamics*. John Wiley & Sons.
8. Bose, A., Thomas, I., & Abraham, E. (2018). Fluorescence spectroscopy and its applications: A Review. *Int. J. Adv. Pharm. Res, 8*(1), 1–8.
9. Calamini, B., & I Morimoto, R. (2012). Protein homeostasis as a therapeutic target for diseases of protein conformation. *Current topics in Medicinal Chemistry, 12*(22), 2623–2640.
10. Danoff, E. J., & Fleming, K. G. (2011). The soluble, periplasmic domain of OmpA folds as an independent unit and displays chaperone activity

by reducing the self-association propensity of the unfolded OmpA transmembrane β-barrel. *Biophysical Chemistry, 159*(1), 194–204.

11. Dumas, P., Ennifar, E., Da Veiga, C., Bec, G., Palau, W., Di Primo, C., ... & Rial, J. (2016). Extending ITC to Kinetics with kinITC. In *Methods in Enzymology* (Vol. 567, pp. 157–180). Academic Press.
12. Eliuk, S., & Makarov, A. (2015). Evolution of orbitrap mass spectrometry instrumentation. *Annual review of analytical chemistry, 8*, 61–80.
13. Fane, A. G., Wang, R., & Hu, M. X. (2015). Synthetic membranes for water purification: status and future. *Angewandte Chemie International Edition, 54*(11), 3368–3386.
14. Fasman, G. D. (Ed.). (2012). *Prediction of protein structure and the principles of protein conformation*. Springer Science & Business Media.
15. Fermi, E. (2012). *Thermodynamics*. Courier Corporation. *14*(5), 78–82.
16. Gao, K., Oerlemans, R., & Groves, M. R. (2020). Theory and applications of differential scanning fluorimetry in early-stage drug discovery. *Biophysical reviews, 12*(1), 85–104.
17. Gouverneur, V. (2012). *Fluorine in Pharmaceutical and Medicinal Chemistry: From Biophysical Aspects to Clinical Applications*. World Scientific.
18. Hu, J. H., Sun, Y., Qi, J., Pei, P. X., Lin, Q., & Zhang, Y. M. (2016). A colorimetric and "turn-on" fluorimetric chemosensor for the selective detection of cyanide and its application in food samples. *RSC Advances, 6*(102), 100401–100406.
19. Jusniar, J., Effendi, E., Endang, B., & SUTRISNO, S. (2018). The misconception of stoichiometry and its impact on the chemical equilibrium. *Advances in Social Science, Education and Humanities Research (ASSEHR), 227*.
20. Kupitz, C., Olmos, J. L., Holl, M., Tremblay, L., Pande, K., Pandey, S., ... & Schmidt, M. (2017). Structural enzymology using X-ray free electron lasers. *Structural Dynamics, 4*(4).
21. Lawrence, P. B., & Price, J. L. (2016). How PEGylation influences protein conformational stability. *Current Opinion in Chemical Biology, 34*, 88–94.
22. Leney, A. C., & Heck, A. J. (2016). Native mass spectrometry: what is in the name?. *Journal of the American Society for Mass Spectrometry, 28*(1), 5–13.
23. Lewis, G. N., Randall, M., Pitzer, K. S., & Van Zeggeren, F., & Storey, S. H. (2011). *The Computation of Chemical Equilibria*. Cambridge University Press. *Thermodynamics*. Courier Dover Publications.
24. Lima, E. C., Hosseini-Bandegharaei, A., Moreno-Piraján, J. C., & Anastopoulos, I. (2019). A critical review of the estimation of the

thermodynamic parameters on adsorption equilibria. Wrong use of equilibrium constant in the Van't Hoof equation for calculation of thermodynamic parameters of adsorption. *Journal of Molecular Liquids, 273*, 425–434.

25. Macha, M., Marion, S., Nandigana, V. V., & Radenovic, A. (2019). 2D materials as an emerging platform for nanopore-based power generation. *Nature Reviews Materials, 4*(9), 588–605.
26. Martens, L., Chambers, M., Sturm, M., Kessner, D., Levander, F., Shofstahl, J., ... & Deutsch, E. W. (2011). mzML—a community standard for mass spectrometry data. *Molecular & Cellular Proteomics, 10*(1).
27. Moran, M. J., Shapiro, H. N., Boettner, D. D., & Bailey, M. B. (2010). *Fundamentals of Engineering Thermodynamics*. John Wiley & Sons.
28. Mu, Q., Jiang, G., Chen, L., Zhou, H., Fourches, D., Tropsha, A., & Yan, B. (2014). Chemical basis of interactions between engineered nanoparticles and biological systems. *Chemical Reviews, 114*(15), 7740–7781.
29. Parrondo, J. M., Horowitz, J. M., & Sagawa, T. (2015). Thermodynamics of information. *Nature Physics, 11*(2), 131–139.
30. Punekar, N. S. (2018). *Enzymes: Catalysis, Kinetics and Mechanisms*. Springer.
31. Ramanan, V., Siddaiah, B., Raji, K., & Ramamurthy, P. (2018). Green synthesis of multifunctionalized, nitrogen-doped, highly fluorescent carbon dots from waste expanded polystyrene and its application in the fluorimetric detection of Au3+ ions in aqueous media. *ACS Sustainable Chemistry & Engineering, 6*(2), 1627–1638.
32. Reiner, J. E., Balijepalli, A., Robertson, J. W., Campbell, J., Suehle, J., & Kasianowicz, J. J. (2012). Disease detection and management via single nanopore-based sensors. *Chemical Reviews, 112*(12), 6431–6451.
33. Renaud, J. P., Chung, C. W., Danielson, U. H., Egner, U., Hennig, M., Hubbard, R. E., & Nar, H. (2016). Biophysics in drug discovery: impact, challenges and opportunities. *Nature Reviews Drug Discovery, 15*(10), 679–698.
34. Ridolfi, F., Renzulli, A., & Puerini, M. (2010). Stability and chemical equilibrium of amphibole in calc-alkaline magmas: an overview, new thermobarometric formulations and application to subduction-related volcanoes. *Contributions to Mineralogy and Petrology, 160*, 45–66.
35. Rodrigues, C. H., Pires, D. E., & Ascher, D. B. (2018). DynaMut: predicting the impact of mutations on protein conformation, flexibility and stability. *Nucleic Acids Research, 46*(W1), W350-W355.

36. Sauer, S., & Kliem, M. (2010). Mass spectrometry tools for the classification and identification of bacteria. *Nature Reviews Microbiology*, *8*(1), 74–82.
37. Shen, H., Tauzin, L. J., Baiyasi, R., Wang, W., Moringo, N., Shuang, B., & Landes, C. F. (2017). Single particle tracking: from theory to biophysical applications. *Chemical Reviews*, *117*(11), 7331–7376.
38. Silbey, R. J., Alberty, R. A., Papadantonakis, G. A., & Bawendi, M. G. (2022). *Physical Chemistry*. John Wiley & Sons.
39. Srinivasan, B. (2021). Words of advice: teaching enzyme kinetics. *The FEBS journal*, *288*(7), 2068–2083.
40. Stark, W. J. (2011). Nanoparticles in biological systems. *Angewandte Chemie International Edition*, *50*(6), 1242–1258.
41. Stumm, W., & Morgan, J. J. (2012). *Aquatic Chemistry: Chemical Equilibria and Rates in Natural Waters*. John Wiley & Sons.
42. Timp, W., Comer, J., & Aksimentiev, A. (2012). DNA base-calling from a nanopore using a Viterbi algorithm. *Biophysical Journal*, *102*(10), L37-L39.
43. Van Zeggeren, F., & Storey, S. H. (2011). *The computation of Chemical Equilibria*. Cambridge University Press.
44. Waduge, P., Hu, R., Bandarkar, P., Yamazaki, H., Cressiot, B., Zhao, Q., ... & Wanunu, M. (2017). Nanopore-based measurements of protein size, fluctuations and conformational changes. *ACS Nano*, *11*(6), 5706–5716.
45. Whitcombe, M. J., Kirsch, N., & Nicholls, I. A. (2014). Molecular imprinting science and technology: a survey of the literature for the years 2004–2011. *Journal of Molecular Recognition*, *27*(6), 297–401.
46. Wu, E. L., Cheng, X., Jo, S., Rui, H., Song, K. C., Dávila-Contreras, E. M., & Im, W. (2014). *CHARMM - GUI Membrane Builder Toward Realistic Biological Membrane Simulations*. *4*(7), 32–38.
47. Xue, L., Yamazaki, H., Ren, R., Wanunu, M., Ivanov, A. P., & Edel, J. B. (2020). Solid-state nanopore sensors. *Nature Reviews Materials*, *5*(12), 931–951.
48. Yao, Z., Zhang, Y., & Wu, H. (2019). Regulation of C-reactive protein conformation in inflammation. *Inflammation Research*, *68*, 815–823.
49. Zhou, Y., Huang, X., Hu, X., Tong, W., Leng, Y., & Xiong, Y. (2021). Recent advances in colorimetry/fluorimetry-based dual-modal sensing technologies. *Biosensors and Bioelectronics*, *190*, 113386.
50. Zubarev, R. A., & Makarov, A. (2013). *Orbitrap Mass Spectrometry*. *4*(10), 80–86.

INDEX

B

C

D

E

F

G

V

W

X